UNDERSTANDING MARINE ECOLOGY

UNDERSTANDING MARINE ECOLOGY

By

Dr. Ashok Kumar

Deptt. of Zoology
Bundelkhand University
Campus Department
Jhansi

DISCOVERY PUBLISHING HOUSE PVT. LTD.
NEW DELHI-110 002

Published by:
Tilak Wasan
DISCOVERY PUBLISHING HOUSE PVT. LTD.
4831/24, Ansari Road, Prahlad Street
Darya Ganj, New Delhi-110002 (India)
Phone: +91-11-23279245, 43764432
Fax: +91-11-23253475
E-mail: parul.wasan@gmail.com
info@discoverypublishinggroup.com
discoverypublishinghouse@gmail.com
web: www.discoverypublishinggroup.com

First Edition: **2011**
ISBN: 978-81-8356-860-9

Understanding Marine Ecology

Printed at:
Shree Balaji Art Press
Delhi

PREFACE

The present title *"Understanding Marine Ecology"* provides a structured approach to learning by covering all the important topics in a uniform, systematic format. The book has been comprehensively designed incorporating recent advances in this fast moving field. It is written to provide accessible information on marine ecology in compact form for undergraduate students in biology and related life sciences. It will be useful for both beginning students and those who are more advanced. In addition, busy lecturers who require a quick reference compendium will find it useful, particularly for tutional planning. Simple, yet hopefully clear figures and tables are provided throughout the book.

The over-riding goal of this book, and indeed of the whole *Understanding series,* is to present the essential information concering microbiology in a compact, readily accessible form which leads itself to student learning and revision. The convergence of various approaches has generated a rich panorama of detail, the significance of which we are still attempting to unraval. The present text has been written as an introduction to this rapidly growing field.

To make the work more comprehensive and informative, the author has consulted many authoritative books, research journals, abstracts, monographs etc., so there can be no claim to originality except in the manner of treatment.

The author expresses his thanks to his friends and colleagues whose continue inspirations have initiated him to bring out this book.

The author expresses his gratitude to Mr. Wasan and staff of M/s Discovery Publishing House Pvt. Ltd. for their whole hearted co-operation in the publication of this book.

In the mean time, the author will remain sincerely responsible for any shortcomings of the book and be grateful to the readers for their suggestions and constructive criticism for the continuous betterment of the book. He takes this opportunity to appeal to the readers to send their suggestions straightaway to his Publisher.

Author

CONTENTS

Chapter 1

INTRODUCTION

Man is an animal with *ecological* requirements just like any other. Human beings have physical limits of tolerance and optimum levels for survival; they also need a source of *energy* and *mineral* nutrients.

But humans are *peculiar* in that they have the capacity to modify global ecosystems on a very considerable scale in order to redirect the products of those systems into the needs human society.

No other organism has produced such an impact on the physical, chemical and *biological constitution* of the earth. For this reason it is necessary to look carefully at the relationship between man and nature through the eyes of the *ecologist*.

In this way it becomes possible to locate those areas in which the stresses created by human beings are becoming *intolerable* and are threatening the future of certain species and *habitats*. It is also the most hopeful means of *predicting* the outcome of some of these harmful activities and may supply the answers to a number of the issues which are creating the current *environmental dilemma*.

POLLUTION: WHAT, WHERE AND WHY?

Today, the cry of "*pollution*" is heard from all the nooks

and corners of the globe, and pollution has become a major threat to the very existence of mankind on this earth. It is the major *challenge* of our times.

The pollution of various resources has gone to such an extent that we are unable to breath fresh air and drink fresh water. On one hand the *advancements* of Science and *Technology* have added to the human comforts by giving us *automobiles, electrical appliances, supersonic jets, space crafts, better medicines, better chemicals* to control harmful insects and other *pests*, etc., but on the other hand, they have given us a very serious problem to face pollution.

In fact, it is not a new problem, except in *dimensions* which we see today. When man and other higher animals began theft life on this earth, there was absolutely no sign of pollution.

There was perfect balance in various natural processes. The air and water were pure and soil was fertile. The problem of pollution arose with the very *civilization* of man. As soon as man learned to use fire, the air pollution began.

As human pollution increased, there was increase in our working sphere and with this also increased the pollution. The rapid unplanned *industrial progress* added to our pollution problem.

What is Pollution?

Pollution can be defined as an undesirable change in the *physical*, *chemical*, or *biological* characteristics of the *air*, *water*, or *land* that can harmfully affect health, survival, or activities of humans or other living organisms. Note that according to this definition pollution does not have to cause *physical harm*.

It may merely interfere with human activities. For example, a lake may be considered polluted if it cannot be used for boating activities. The problem with defining pollution is specifying what constitutes "*undesirable change*" which requires value judgments.

For example, chemicals spewed into the air or water from an industrial plant may be harmful to humans and other organisms living near the plant. However, if expensive pollution controls are required, the plant may be forced to shut down.

Thus, workers who would lose their jobs may feel that the risks to them from *contaminated* air and water are not as serious as the benefits of having jobs. Value judgments about short-term versus long-term risks can also cause *controversy*.

Building a world wide network of nuclear power plants may provide electricity for the present human generation, but it compels future generations to handle and store the *radioactive* wastes produced for thousands of years, even if nuclear power is abandoned or exhausted as an energy source. Thus, we see that the determination of desirable versus *undersirable* effects of an *environmentally* altering activity is a very difficult and highly *controversial* process.

The nature of tragedy, as the philosopher Hegel pcinted out long ago, is the conflict not between right and wrong but between right and right, Nevertheless, although defining pollution is a difficult, *controversial* process, it must be done in order to control pollution.

Types of Pollution

From a biological viewpoint, we can recognize two major types of pollutants: *degradable* and *non-degradable*. A degradable pollutant can be decomposed, removed, or consumed and thus reduced to acceptable levels either by natural processes or by human *engineered* systems (such as *sewage* trestment plants), as long as the systems are not overloaded.

There are two classes of degradable pollutants rapidly degradable (non-persistent) and slowly degradable (*persistent*). Rapidly *degradable* pollutants, such as human sawage and animal and crop wastes, can normally be decomposed rather quickly if the system is not overloaded.

For example, a rapidly flowing river can normally

cleanse itself of some human sewage fairly quickly, but this natural cleansing process can be overwhelmed by too much raw sewage from a large city or a number of small *cities* or farms. Slowly degradable *pollutants*, such as DDT and some radioactive materials, *decompose* slowly but eventually are either broken down completely or reduced to *harmless* levels.

Some radioactive materials that give of harmful radiation, such as iodine-131, decay to harmless levels in a few minutes or hours and are rapidly degradable *pollutants*.

Others, such as strontium-90 (produced by nuclear bomb blasts) ; can persist at harmful levels for decades ; *plutonium-239*, produced by nuclear power plants, can remain at harmful levels for many thousands of years.

Slowly degradable pollutants are often *synthetic* compounds, such as *DDT* and *plastics*. Because they are *artificial* substances, nature usually bad not evolved processes for breaking them down.

Potentially harmful pollutants that break down slowly must either be prevented from reaching the environment or controlled so that they do not build up to harmful levels. Non-degradable pollutants are not broken down by natural processes.

Examples of *non-degradable* pollutants are *mercury*, *lead*, and some of their compounds and some plastics. Like slowly degradable pollutants, non-degradable pollutants must be either prevented from entering the air, water, and soil or kept below harmful levels by *removal* from the environment.

Air Pollution

Air is the most vital component, without which the question survival does not arise beyond a few minutes. A human being of normal health and height respires 22,000 times a day and the air utilized in course of such respiration is about 20 times more than the quantity of food consumed by him.

Natural air at mean sea level contains *nitrogen* (78.09%), *oxygen* (20.94%) and *argon* (0.93%) as *macro-constituents*. The microcomponents of air carbon dioxide (0.0318%), krypton (0.0001 %) and helium (0.00052%). Many other compounds *e g.,* SO_2, NO_2, O_2 etc. are also present in much lower *concentration* than those mentioned.

Besides these, water vapour and dust *particulates*, in varying amounts and sizes, are, present in air, depending upgn the *meteorological*, topographical and industrial activities of the areas under examination. *Current* of air is known as wind blowing with velocity *ranging* from I to 6 km/hr is known as light air.

Air with velocity of 7 to 12 km/hr known as light-breeze and those from 19 to 44 km/hr as strong breeze. Air current exceeding the velocity of 90 km/hr is called storm and over 104 km/hr is known as hurricane. Such *abnormal* condition in air movement *releases* too much of dust particulates into the *atmospheric* air.

Estimates have shown that, even now, hardly 12 to 15% of the total dust particulates released into the environment is manmade and the rest is by *nature*. Though man-made air pollution is statistically insignificant, it *nevertheless*, *poses*, *great danger*, because such pollution usually contains toxic elements as opposed to natural pollution.

The second reason is that industrial air pollution is often around human habitation resulting in the living population being badly affected. The pollution caused by vehicular traffic is maximum in the heart of the urbanized areas, resulting in the constant exposure of the urbanites to lead, nitrogen oxide and unburnt particles of hydrocarbon.

These released pollutants subsequently turn into the photo chemically active species e.g., ozone, PAN etc. which often produces smogy environment especially during winter. Air prevails even upto 100 km above the surface of the earth and 11 km below, at the deepest point in the Pacific ocean around *Phillipines*.

The pressure of air drops from one atmosphere at sea level to 3×10^{-7} atmosphere in the mesophere region. The total mass of the atmosphere is about 5×10^{-6} tons which is roughly one millionth part of the earth's total mass 80% of the total amount of the *atmospheric* air is confined in the troposphere. within 12 kms above tle mean sea level.

Meteorological Factors Responsible for Air Pollution

There are three major meteorological factors which govern the dispersal pollutants from the source into the surrounding atmosphere.

(i) Wind profiles are caused by the *interactions* of forces due to pressure, the earth's rotation and friction. On a *windless* day, the wind speed is zero at the earth's rotation and friction.

On a *windless* day, the wind speed is zero at the earth's surface but gradually increases to a value representative of free atmosphere at a height which is of the order of 1 km, known as '*planetary boundry layer*'.

(ii) The temperature is another profile which significantly affects the dispersion of pollutants. Super-adiabatic condition indicates unstable atmosphere which is highly desirable as it causes rapid dispersion of pollutants.

Sub-adiabatic as well as inversion conditions are highly dangerous, leading to the accumulation of the pollutants and ultimately causing severe air pollution problems in the area.

(iii) Topography often influences the local wind, as for example, the mountain valley wind, where the air tends to flow down the valley at night. Similarly, the *sea breeze* in the coastal regions causes the stack *effluents* to drift over land during day and no pollution *apparently* appears in the night.

In order to fully understand the mechanism of air pollution in the context of the roles played by meteorological factors, let us recall the different layers of the atmosphere.

Of special significance are the trochosphere and troposphere regions of the *atmosphere* to fully appreciate the reasons of the air pollution *phenomenon*.

Trochosphere is the region where cloud formation takes place and generally extends upto 1.5 km. The height extending upto 15 km from the surface of the earth is covered by troposphere where more than 80% of the total air of the *biosphere* remains confined.

Within this range, the air temperature decreases with increasing altitude. This decrease in temperature with altitude is called adiabatic (no heat *transfer*) lapse rate and is equal to six degree centigrade for every 1000 metre height. When the temp-erature of air decreases faster than the *adiabatic lapse* rate (a condition known as *superadiabatic* lapse rate), then the atmosphere is considered to be unstable.

As such, the air from one altitude readily mixes with air from other altitudes which *ultimately* leads to the formation of *looping* type of *plume*.

A *fanning* type of plume is caused by *thermal* inversion where vertical movement of pollutants is limited. The fumigation type of plume results when there is inversion above the plume and instability below. Such condition occurs when temperature instead of decreasing, increases with *altitude*, thus forming a very stable *atmosphere*.

The fumigation type of *plume* acts as lid over the atmosphere and brings down the pollutants to ground. Thus, creating severe problem of air as well as soil and water pollution. Pollutants in air are likely to accumulate in concentration sufficiently large enough to cause such episode in the regions where several days of low *mixing height*, low wind speed and no precipitation occur.

The combination of these circumstances often results in air *pollution episode*. Studies carried out by scientists reveal that mixing height of less than 1500 metre, wind speed of less than 15 kmph (4 m/sec) and insignificant precipitation for about 4 days are very favourable conditions for creating air pollution.

Types and Sources of Air Pollution

Air pollutant are generally of two types.

(i) Primary pollutants are those which are directly produced as a result of interaction between two or more primary pollutants *e.g. nitrogen peroxide, ozone. sulphur trioxide, aldehye, peroxyacetyl nitrate* (PAN), etc. Secondary pollutants are *photochemically* active and responsible for the formation for smog.

Smog is a combination of smoke and fog. Smoke is an emulsion of unburnt carbon particles in air and fog is the emulsion of water vapor in air. Particulates present in the *atmosphere* may vary in numbers from several hundred/ cm^3 in very pure air to more than $10^3/cm^3$. Particulate mass level ranges from 10 μgm^3 in clean air to 60-2000 $\mu g/ms$ in the polluted air in urban areas.

The particles of size between 0.1-lu govern the heat balance, fog formation and provide nuclei for ice and water *droplets formation*. Smog has now a days become a general phenomenon in big cities, especially during winter, primarily caused by automobile *exhaust*. The *automobile* exhaust contains *unburnt hydrocarbon* which produces *photochemically* active compounds known as *formaldehyde* and causes eye *irritation*.

Lead is another air pollutant, generated from the automobile exhaust, owing to use of *tetraethyllead* as *antiknock* material in the petrol. Mosses are excellent indicators lead pollution from aerial sources. The lead *poisoning* causes elevated content of amino leveulinic acid (ALA) in the *urine* of the *affected person*.

The increased concentration of bromine in the water or soil is another indirect indication of lead *contamination*. *Evidences* have shown that consumption of dust and dirt, eomposed of lead outfall is a major source of unusual *indigestion* in urban people. *Dust* and *smoke* are commonly found in air.

It is the respirable dust index (RDI) *i.e.* size of dust particulates, which affect the human health. Particulates of size 3 micron (u) reach the lungs, out of the total *inhaled dust*, while particulates bigger than this size are chocked somewhere in the *respiratory* passage of nose. Particulates of much lower size 0.1 (u) are removed through expelling *mechanism* of the *human body* as cough.

Out of the total retained *particulates*, still some get dissolved in the system. Thus, being eliminated without showing any harmful effects. Even all the permanently retained particulates are not harmful ; some of them are deposited indefinitely in inter state on the lungs walls without any *detectable injury*.

Only reactive deposited particulates cause harmful effects, but they are not equally harmful. The *respirable* dust *production* ranges from 27 to 340 gm for one ton of coal mined. In addition to the automobile exhaust, other major sources of air *pollution* are mining, *thermal power plants* and *brick kilns*.

The latter is a pin-point source of air pollution, especially in the rural area. The intensive coal mining regions of *Jharia*, *Dhanbad*, *Katras* and the adjoining areas of Bihar have been identified to be the most air polluted cities nearly 15,000 tons per year of particulate matters are released in the air in Jharia coal field areas, causing complete *blanketting* effect when viewed aerially.

Similarly, open stack coal carbonization emitts about 35,000 tons of particulate matters annually in the atmosphere. The release of sulphur dioxide through burning of coal in the boilers in thermal power stations is causing severe damage to the plants, concrete structures and also health hazard, owing to the oxidation of SO_2 into sulphuric acid. Studies conducted by Central Mining Research Station at Dhanbad (1986) indicate the mean concentration of Fe (2.48 ug/m^3), (0.55 ug/ma), Ni(0.04-0.57 ug/m^3), Cu(0.05-1 ug/m8), *Cr* (max. 0.55 ug/ma) Zn (max. 2..8 ug/m^3) in ambient air in the Jharia coal field areas.

Effects of Air Pollution on Our Cultural Assets

Neon in excess of 18 ppm, helium greater than 5.2 ppm, *methane* higher than 1.4 ppm, *krypton* , more than 1.0 ppm hydrogen in excess of 0.5 ppm, nitrogen dioxide more than 0.001 ppm, *carbon monoxide* in excess of 0.1 ppm, and *sulphur dioxide* greater than 0.002 ppm in air injurious for human health. Ozone (O_3) in concentration of over 0.3 ppm in air causes nose and eye irritation.

Its presence at a level between 1 to 3 ppm causes extreme fatigue after 2 hours of exposure and at higher than 9 ppm level, causes pulmonary edem (a discase in which excessive fluid occurs) in the lungs. Ozone concentration in 0.1 ppm for 20 days exposure causes 50% reduction in pollen germination in case of tobacco plant.

Exposure of plants to ozono causes yellow spot formation on leaves which subsequently affects the rate of photosynthesis considerably.

Exposure to ozone at concentrations of 0.02 ppm for 65 minutes, 0.26 ppm for 5 minutes, ppm for 3 minutes, and 20 ppm for 1 second produces cracks in rubber.

PAN and aidehyde found in smog are eye irritants. Exposure to carbon-monoxide (CO) at concentration of 100 ppm even for an hour may be fatal for human kind, probably because of its affinity with haemoglobin (red corpuscles of blood) which is about 200 times more as compared to oxygen.

Sulphur dioxide (SO_2) is one of the most potential pollutants present in the air over and around industrial areas materials like *papers*, *textiles*, *leathers*, *stones*, *marbles*, *paintings* and *metals* are severely attacked by sulphur dioxide in presence of atmospheric *moisture*.

The conversion of atmospheric sulphur dioxide into sulphuric acid is known as *sulphonation.* The process of sulphonation is accelerated in the presence of *moisture* and iron. Iron is known to be present in dust, nails etc. and moisture is present in the material itself.

Cellulose which is the basic building material of paper, cotton and linen is rapidly decomposed by the process of hydrolysis. As a rest It, the paper and textile goods become *yellowish, brittle,* and ultimately loose their strength and *fibrous* structure.

Similarly, Collagen (which is the basic constituent of leather) is also disintegrated by SO_2 and thereby, the leather becomes powder y in due course of time. Stones particularly calcarious, like marble, which are essentially *calcium carbonate,* are severely damaged by *sulphuric acid.*

Calcium carbonate is converted into calcium sulphate, resulting in huge incrustation by which the beauty of marble is changed. Paintings in which lime plaster is used, become expanded in volume and look flaky and powdery, owine to their convension into the corresponding sulphate under the action of the atmospheric sulphur dioxide.

Acidity generated by atmospheric sulphur dioxide initiates the electrolytic corrosion of metal. The over all effect of SO_2 is summarised as belowL

(i) 1 ppm for 1 hr or 0.2 ppm for 8 hrs causes plant damage.

(ii) 5 ppm for 2 hrs is a high level of concentration which causes broncho-constriction in human body.

(iii) 10 ppm for 1 hr exposure is considered a distress level for human beings.

Objects made of coper and silver are much more sensitive towards hydrogen sulphide, even in much lower concentration as compared that of sulphur dioxide. Both the metals are converted into their respective sulphides and consequently loose their artistic, ornamental and antique values.

Numerous monuments and ancient temples are exposed

to the sea winds, coupled with chloride impregnated fogs at the coastal sides.

The chloride ions absorbed by the stones during *foggy weather* are carried to the surface of the stones in dry weather, causing efforesene/ce and finally leading to the process of complete *distintegration* of the entity.

Bronze, an alloy of copper, zinc and silver, which constitutes a large part of artifacts, is badly corroded by sea winds. The most probable mechanism of the bronze corrosion may be presented as such. The chloride ions react with copper and form unstable cuprous chloride.

The latter farther reacts with environmental oxygen and moisture forming cuprous oxide, cupric chloride and hydrochloric acid. Thus, a set of chain reactions finally reflects upon the bronze objects which ultimately *destroy* them.

Ozone (O_3) is highly sensitive towards saturated organic materials, even at room temperature and causes their *depolymerization*. Resins, varnishes etc. are used in paintings and are products of *saturated organic* compounds.

Woollen, cotton, paper and all other such materials that consists of cellulose are also liable to be photo-oxidised by ozone and colours.

The bloom in painting is caused by environmental ammonia (NH,) in concentration sharply detactable by human nose, due to the formation of a compound known as a *ammonium sulphate*.

Lower *concentration* of ammonium is though, useful because it takes care of free atmospheric acidity and prevents other damages.

Cellulose, showing amphoteric behaviour, is *hydrolysed* and *oxidised* under both acidic and *alkaline* condition. As such, the materials rich in cellulosic content are also at constant risk of being degraded in higher concentration of environmental ammonia.

Furthermore, it inflames the upper respiratory passage, even in low concentration, under longer exposure. *Aldehyde*

(CH_3CHO) irritates nasal and *respiratory* tract, *arsines* breakdown the red cells in the blood, chlorine attacks entire respiratory tract and mucous membrane of eyes, *hydrogen cyanide* interferes with nerve cells, *hydrogen fluoride* irritates and corrodes all body passages, possibly leading to cancer.

Air Pollution Control

The safety of three M's — Maa, material and machinery must be the ultimate object behind the measures to control the environmental pollution problems.

Pollution control measures should be considered as socia' responsibility and obligation not only by the concerned departments, but also by all who have faith in the need for preserving the valuable natural environmental resources.

The implementation of the control measures should be based on the principle of recovery or recycling of the pollutants and must be taken as an integral part of production *i.e.*, never as a liability but always as an asset. Air pollution may be controlled by physical as well as by chemical methods.

Some of the most effective methods are shown below:

Dilution Method

Physical methods include the suitable height and design of the stack. The minimum height of the stack has been worked out to be 38 metres in order to ensure proper mixing of the discharged *pollutants* into the adjoining atmosphere.

Air Cleaning

Air cleaning device removes or recovers sbspended solids, liquid and also soluble gases and vapours from air before they are discharged into the atmosphere. The commonly used air cleaning devices are :

(1) Gravity settlers,

(2) Cyclone separator,

(3) Wet collectors, and

(4) Electrostatic precipitators.

Substitution Method

The method is based on the principle to substitute the offending materials and processed by less offending ones.

Plantation

Intensive plantation in the region considerably helps in reducing dust, smoke and other pollutants. Studies conducted by the authors indicate that the trees with simple leaf, i *e., pipal*, *pakur*, *banyan*, *teak*, *sal*, *debdaru*, *mango*, *kachnar*, *kadomba* etc. are more effective in removing dust particles suspended in the air than those of the trees *e.g. ashoka*, *palas*, *nim*, *sandal* and *gulmohar* with compound leaf.

The study further shows that the dust removal from the air is best achieved by banyan (5.35+0.12 g/m^2 of leaf), followed by debdaru (4.56 g/m^2), sal (4.50 g,/m^2, pipal (4.10 g/m2) pakur (4.09 g/m^2) and the minimum by *kadomba* (3.57 g/m^2) among the trees with simple *leaf*.

Similarly, the highest proportion of dust particles settlement from the air is ensured by *ashoka* (3.78 g/m^2) followed be *palas* (3.06 g/m^2), nim (2.92 g/m^2), *sandal* (2.24 g/m^2) and the least is achieved by *gulmohar* (1.44 g/m^2) among the trees with compound leaf.

Therefore, along the road side and at places where maximum dust generation is possible, trees of simple leaf should preferably be planted. For ensuring the involvement participation of the public, *plantation* should be encouraged on the pattern of the social forestry.

The principle of atleast one tree for each school going student, two by each teacher and three during rest of the period of life by each individual should be practised in true sense of the term.

Besides, the Forest Conservation Act of 1980 should be updated and be made stringent for those who are cruel to trees and plants.

Chemical Methods

This method is usually for removing sulphur dioxide

from air, resulting from the burning of coal, with rich sulphur content. Several chemical methods have by now been developed to have an effective control over air pollution problems. Briefly, some of these methods are discussed below:

Beneficlation Process

By this method coal is treated with lime or dolomite and processed in such a way that both the sulphur and ash content of the original coal are significantly reduced with improved yield of energy on burning. Thus, the release of sulphur dioxide is automatically reduced considerably. The other method is to fix the sulphur dioxide in the combustion chamber itself by the *spraying* of ammonical liquor.

Reduction Process

The SO_2 emission is absorbed by alkaline alumina, resulting in quick reduction of the former into the elemental sulphur.

Catalytic Oxidation Process

Vanadium oxide is generally used as catalyst where by SO_2 is converted into H_2SO_4 or ammonium sulphate.

Control Measures for Smog

Four techniques are generally used to control hydrocarbon emission from stationary sources. They are incineration, adsorption, absorption and condensation of the exhaust gases. In the *adsorption* method, exhaust gases are passed through a bed of a granulated adsorber where hydrocarbon vapours are adsorbed into the adsorber.

The absorption method is similar to adsorption, except that the exhaust is brought into intimate contact with a liquid in which the hydrocarbons *dissolye* or become *suspended*. In the condensation method, exhaust gases are passed over low temperature surfaces where they are condensed.

Liquid hydrocarbons are left behind and collected. Control of hydrocarbon emissions from automobiles is more complicated. Several engine design characteristics favour lower exhaust of hydrocarbon emission. For example, reduction of surface/volume ratio of the combustion chamber decreases *hydrocarbon* emission markedly. Ill maintained auto-vehicles with out-dated *silencers* should not be allowed to play on the road.

WATER POLLUTION

Water is relatively more prone to pollution than air because it is more viscous, having lesser moment of inertia and possessing closer molecular distance than air. Air, being available in the environment in much larger proportion as compared to water, the pollutants in air are quickly *diluted* and dispersed away along the down wind direction. Pollution load is thus, *drastically* reduced in air. The unequal distribution of water on the surface of earth and the fast *declining* availability of fresh usuable water per capita per year, due to increasing population, are the major causes for concern in terms of the quantity and quality of water.

Water pollution problem is more delicate and severe than that of air. The fundamental difference between the pollution of air and water is as such. Air pollution is of widespread nature whereas water pollution remains more or less confined.

Therefore, the latter is localised in nature. Air pollutants are known to travel thousand of kilometres without loosing their identity and, therefore, are not matters of immediate concern. The pollutants in water undergo many reactions and subsequently sometimes turn into hazardous forms.

The study of air pollution is costly but the principles involved are relatively simple than in water pollution studies and management. Water may be present in rivers, ponds, lakes and in ground reservoirs. The surface water bodies get easily polluted whereas ground water reservoirs become polluted only after prolonged exposure.

Among the surface water bodies, the rivers are less prone to pollution than ponds and lakes, because of the current available in the former. But once the rivers become polluted, complex and costly techniques for control of such pollution are needed.

The same is true in case of ground water pollution, which is difficult to treat when and where such problems exist. So far as the effects of pollution on the living biota are concerned, they are again of conflicting nature.

The rivers disperse away the pollutants immediately in the downstream and thus, become free from pollution, but the effects of pollution become established and detectable more easily in river than in ponds and lakes because the biota in river and generally less tolerant to pollution whereas the biota living in the stagnant water bodies are more tolerant to pollution.

Sources of Water Pollution

There are generally two sources of water pollution

(i) *Point sources* of pollution are of definite identity with almost constant volume and fixed composition of the effluents discharged. Industrial waste waters are examples of the effluents generated from point sources of pollution.

(ii) *Diffused sources* of pollution are points, situated far beyond the sight having wastes of unknown composition and volume. The street run-off, agricultural run-off and other such wastes are examples of the effluents produced from the diffused source of pollution.

It is much easier to design and operate the treatment plants for correcting the *effluents* produced from the point sources of pollution than for the effluents produced from the diffused sources of pollution.

It is practically impossible to treat the effluent of the latter because of very huge investment of money, time, labour and the lack of availability of *sophisticated* technologies in our country. The potential source of water pollution in all developing or underdeveloped countries are *sewage* and *industrial wastes.*

For example, out of the total wastes discharged into river Ganga, volume of sewage discharged constitutes 72 per cent whereas 25 per cent discharge is contributed by industrial wastes. In term of BOD, the pollution load contributed to river *Oanga* by both the pollutants is in 1 : 1 ratio.

The effluents from agricultural and street run-offs join the river in much larger proportion, especially during the monsoon. The excreta contributed by *nomadic* people living along the banks of rivers or in the respective catchment areas are the only potential diffused source for causing the *bacteriological* pollution of surface water bodies in India.

Sewage

Sewage constitutes of 99.9 per cent solid and 0.1 per cent liquid. 70 per cent of these are organic constituents and the remaining 30 per cent are inorganic. 65 per cent of organic portion are proteins while *carbo-hydrates* and fats are representsed by 25 per cent and 10 per cent respectively.

Estimates made by Arcievalla suggest that 6 to 12g of nitrogen and 1 to 3g of phosphorous are generated by every person every day in the form of sewage. Sewage excreted by a human being contains more than 400 different species of bacteria and *viruses*. About 100 viral species in sewage are known to cause diseases in human beings.

Out of this, 90 per cent are contained in human sewage, 100 ml of sewage contains 16^{9}-10^{10} bacterial cells out of which 10^{6}-10^{9} may belong to *coliform* groups and 10^{6}-10^{6} may be of faecal origin. The Salmonella typhosa variety ranges between 10^{1}-10^{4}.

Sewage has been established to contain potential firsh *pathogens* such as *Acromonas*, *Pseudomonas* etc *Myxobacteria* are also present in water, polluted particularly by sewage. Well treated sewage may still contain pathogenic bacteria and viruses, unless properly *chlorinated* before its discharge into any water course.

Sewage, on drying after treatment, is known as sludge. Sewage further diluted by kitchen and other domestic wastes are known as sullage. Sewage may be strong, average and weak. Assessment of its strength is determined on the basis of *chloride* content, the ratio of BOD and permanganate value and on the basis of a *cotimbinaon* of chemical *parameters*, as shown below.

Chloride

Weak sewage contains an average of 60 ppm chloride, sewage of medium strength shows 100 of chloride and strong sewage indicates 150 ppm of chloride.

BOD/PV (3 minutes) Ratio

Weak sewage indicates BOD/PV (permanganate value) ratio equivalent to 60, average sewage near about 100 and strong sewage shows a ratio of over 170.

Potential Pollutants in Industrial Wastes

Industrial effluents generated from sugar, distillery, tannery and paper industries are accompanied by very high organic load with BOD values ranging from 2500 to 1,22,00 ppm. The paper and pulp industrial waste contains Lignin which is made up of *cellulosic material* and remains unattended by microbes for long, causing depletion of fishes up to as long as 40 km distance in the downstream.

The wastes from oil refinery and steel industries contain the most objectionable pollutant known as *phenol*, causing high odour and toxicologic,1 problem in the receiving water body. Besides, the latter also contains cyanide in appreciably high concentration to produce detrimental effects on the *biota* in the receiving water bodies.

Fertilizer industries discharge effluents with high content of *ammonia* (100 ppm), *urea* (180 ppm), *phosphate* (40 ppm) and *sulphate* (1200 ppm), which subsequently find their ways into the adjoining water cources, causing *bloom* of algae.

Ammonia in concentration greater than 0.11 ppm in river, stream or lake is toxic to fishes. The fertilizer waste is usually accompanied by arsenic and chromium and unfortunately both of them are equally objectionable for *aquatic fauna* and *flora*. The wastes generated from *alkali* manufacturing industries contain mercury, which builds up 1000 times higher concentration in fishes than found in the receiving water course.

The minimata disease of Japan is a well known examples of mercury *hazard* where thousands were poisoned and hundreds died by consuming *mercurised* fishes.

Lead is another accumulating poison which is found in wastes generated from *battery*, *printing*, *petroleum* and *pastes* processing indusiries. So many of trace and toxic elements, e.g., *zinc*, *boron*, *copper* etc. are found in large concentration in the *effluents* produced by *nonferrous* industries.

The effluents generated from mining operations are either highly alkaline or acidic and as such both are *injurious* for the aquatic organisms. The phosphate Pyrite mining produces wastes of very high acidic nature, accompanied by as low as pH value as 1. The mining mica and coal sometimes produces acidic wastes but these are not as strong as *pyrite effluent*.

All the industrial wastes generated during processing and manufact-uring stages must be suitably treated before their discharge either on the land or into the water course. The cost involved in their treatment is *negligible* in comparison to the gain achieved.

Tolerance Limit for Pollutants Present in Water

Water is used for a variety of purposes. It may be used for drinking, bathing, cooking, irrigation and industrial purposes. Its industrial use may again be of various types. It is used in *boiler*, *laundary*, *power generation*, *breweries*, *dairies* and in many edible productions and *processings*.

The standard of the quality of water vary considerably in each of the 'above cited industries. Detail for the quality of water standards, applicable in each of these cases is not possible to present here. However, some of the important parameters for the maximum permissible tolerance limits are below:

(i) *Physico-chemical Standard for drinkine water.* Lead, 0.1 ppm, Cadmium and Selen&nm, 0.01 ppm for each, Chromium, Arsenic and Cyanide, 0.05 ppm for each, Phenol and Mercury, 0.C01 ppm for each Turbidity 10 JTU, pH, 6.5 to 9.2.

Total solid, 500-1500 ppm. Hardness ($CaCO_3$): 600 ppm, Magnesium: 150 ppm, Iron: 0.3-1.0 ppm, Manganese-0.5 ppm, Copper-1.5 ppm, Chloride-1000 ppm. Sulphate-400 ppm, Flouride1.5 ppm, Nitrate as (NO_3)-45 ppm, Nitrate as (N)-12 ppm, Zinc-15 ppm.

(ii) *Bacteriological Standard for treated drinking water samples.*

1. E. Coliform should be absent.
2. 95% of the samples examined should not have any coliform count throughout the year.
3. If coliform counts were found, than in no case, it should exceed l0/100 ml.

or

Coliform could should not be found in any two consecutive samples.

(iii) *Maximum permissible chemical limits for inland surface water, intended to be used as raw water for public water supply and bathing ghats.*

BOD-3 ppm, pH-6 to 9, Chloride-600 ppm. Sulphate-100 ppm, Cyanidc-0.01 ppm, Arsenic-0.2 ppm, MPN count-5000/ 100 ml.

(iv) *Bacteriological standard for raw water to be used as source of Public water supply.*

Relaxation has, however been granted in case the total coliform count exceeds the maximum permissible range of 5,000 100 ml for class C water sample, as under.

Not more than 20% of the total number of water samples examined should exceed the maximum permissible range for total coliform (i.e., 5000/100 ml) or almost 5 per cent of the total samples may hardly exceed 4 times the maximum prescribed limit. One more condition had been added by IS, 229-1982, which states that faecal coliform count should not exceed 40 per cent of the total coliform count.

(*v*) *Limits for industrial effluents to be discharged into water course (IS : 2490-1974)*

BOD-30 ppm, COD-250 ppm, Total svspeuded solid (TSS)-100 ppm, Sodium-0.2 per cent, pH-5.5 to 9.0, Temperature40°C. Chloride-600 ppm. Fluoride-2 ppm, Zinc-5 ppm, Copper3 ppm, Nickel-2 ppm, Cadmium-2 ppm, Chromium and Lead 0.1 ppm each, Arsenic and Cyanide-0.2 ppm each, Mercury-0.01 ppm, Selenium-0.05 ppm.

(*vi*) *Limit for industrial waste to be discharged into public sewers (IS: 3306:1974)*

BOD-500 ppm, pH-5.5 to 9 TSS-600 ppm, Temperature45°C, Chloride-600 ppm, Sodium-60 per cent, Cyanide-2 ppm, Copper-3 ppm, Lead-1 ppm.

(*vii*) *Limit for the discharge of sewage into the water course (IS: 4754-1973)*

BOD-20 ppm, TSS-30 ppm.

(*viii*) *Maximum tolerance limit for water intended to be used for irrigation*

The suitability of water to be used for irrigation should strictly satisfy the following criteria

(*a*) Sodium Absorption Ratio (SAR)

$$SAR = \frac{Na}{(Ca + Mg/2)^{1/2}}$$

Note : The values of SAR should not exceed 26.

(*b*) Soluble sodium percentage (SSP)

$$\frac{Na \times 100}{Ca + Mg + Na}$$

Note : SSP must not exceed 60, otherwise the water would become unfit.

(*c*) Potential soil salinity (PSS)

Cl (epm) + ½ SO_4 (epm) where, epm is equivalent part per million which is equal to ppm/ev and ev is equivalent weight.

The water having 6 epm or more chloride is injurious for acidic plants.

(*d*) Residual sodium carbonate (RSC): (CO_3 + HCO_3)(Ca + Mg) in epm. Water with RSC greater than 2.5 is unfit, and those with RSC between 1.25 and 2.5 is marginal and water with less than 1.25 epm of RSC is good.

(*ix*) *Chemical quality criterials of grounwater.* The water exceeding the maximum permissible concentration of any single or a number of constitutents, not only unfit for drinking, rather creates many problems.

For example, waters having more than 400 ppm of bicarbonate, 100 ppm sulphate or 40 ppm silicate may contribute incrustation to the carrying pipes. Waters showing 2 ppm or more of iron or 1 ppm or more of manganese tend to accentuate the process of incrustation, especially when sulphate, silicate and bicarbonate are at optimum concentration in it.

Similarly, waters with low total dissolved solid (TDS) and low PH value contribute to chemical corrosion whereas, the waters with high TDS, usually greater than 1,000 ppm of high pH value lead to the electro-chemical corrosion.

(*x*) *Biological quality criteria for limnological study.*

Biotic Index (BI)

A stream, pond of lake with a biotic index of more than 15 is clean and when it is lower than 10, is considered to be polluted.

Biotic Index = 2 (number of class 1 organisms)+ (number of class II organisms).

Class 1 organisms include caddis fly, may fly etc. and

are pollution intolerent. The organisms belonging to class II category viz. dragon fly, shrimps, plenoria etc. are indicators of mid pollution Class III organisms *e.g.* mosquito larva, chironomoids tubifex etc. are indicators of heavy pollution and their presence marks zero biotic index.

Sequential Comparison Index (SCI)

The determination of the values for SCI does not require much of a background of biological knowledge, rather it is done through simple technique, even by persons of ordinary skill. The mathematical expression for SCI values are as such.

$$SCI = \frac{\text{Number of runs}}{\text{Number of organisms}} \times \text{No. of equivalent species}$$

The SCI values decrease with increasing intensity of pollution because the latter leads to a significant reduction in the number of community species at the cost of the increase in the number of individuals of a particular tolerant species.

Generally, SCI value less than 5 indicates severe pollution, between 5 to 10 mild pollution whereas a value above 15 is indicative of clean water zone. For confidence limit of 90 per cent with mean difference of 100 per cent, requires a minimum of 7 to 8 times determination for each of all the samples drawn from the bottom.

(*xi*) ***Algal Genus Index (AGI).*** Score of 20 or more secured by the algae belonging to different general in water sample is indicative of organic pollution. Score lying between 15 and 19 is taken a probable indication of high organic pollution. AGI value lower than 5 is indicative of clean water zone. The presence of algae means the presence of 50 or more individuals of a single genus in one ml of a water sample.

SOIL POLLUTION

Soil sustains innumerable microbes and large number of plants and animals along with plenty of mineral reserves for purposeful exploitation. Soil has different meaning to different people. Chemists consider the soil like an object endowed significantly with high buffering capacity while to a geologist the soil is a means for ascertaining the age of the earth from a large number of isotopes present in it. The soil to a limnologist is a dependable source of nutrition for the biota, and to a farmer, the soil in the form of land is a source of living whereas to a sanitary engineer, it is nothing more than dumping ground for all kinds of wastes.

Above all, the soil is a living record from where originated all major ecological events of the past. Land is a valuable but fast depleting resource with increasing population. It has been estimated that the total surface area of the earth is 3, 15, 14,640 km^2.

About 30 per cent or 91,56,800 km^2 is lard surface. Of this nearly 80 lakh km^2 forms the Antarctica, which is too cold to be inhabited by human population Another I crore 44 lakh km^2 area is covered by desert and 1 crore 75 lakh km^2 is mountaneous regions (e.g. 80 lakh km^2) are not habitable.

Thus, human population is left with hardly 75.2 m km^2 (7 crore 52 lakh km^2) of land surface to live in. Presently, there are more than 75 people living per km^2 of land. The per capita availability of habitable land is likely to be further reduced to a less than 50 percent of its present proportion by 2000 A. D. (Many of the highly urbanized centres of the world are faced with acute space problem.

For example, there is hardly 20 sq. ft. of living space per capita available in a crowded city like Calcutta, thereby, causing a serious ecological concern. The decreasing availability of living space is not the only problem faced by the world ; rather the fast degradation of land is a more serious aspect of pollution.

India is loosing nearly 6,000 million tons of top soil per year in Ganga basin alone, caused mainly by ruthless deforestation, thus leading to severe erosion problem. Statistics prepared by the Ministry of Agriculture, Government of India (Referred from the World Environment,

1985), indicate that out of 304 million hectare (mh) of land, nearly 175 mh is becoming less and less productive year after year.

Such increasing trend of desei tification is really devastating and needs our urgent attention towards waste land management.

Besides, increased intake of pesticides fertilizer along with radioactive out-fall, is putting additional poisonous ingredients into the land surface and causing serious pollution problem.

Sources of Soil Pollution

There are numerous sources for soil pollution, but the important, among them, are the following three:

(*i*) Indiscriminate discharge of industrial effluents on land and in water course.

(*ii*) Open defecation by animals and human beings especially in rural areas of the country.

(*iii*) Unscientific disposal of solid waters on land.

In developing countries like India, the major source of soil pollution is of the third type, and is prevalent in both urban as well as in rural regions.

The garbage generated by the community in India contains 30-40 per cent of compostable matter from which good qualities of manures can be obtained through anaerobic digestion. Indian refuse, usually contains less than 1 per cent of plastic and glass materials.

The paper content of the refuse varies between 3-7 per cent. Further, among the physical characteristics shown by them, the density of the refuse varies from 330-560 kg/m2 and the per capita refuse production in Indian cities varies from 0.15 to 0.35 kg/day, depending upon the social, economic and educational status of the people living in the community.

Generally, the people with modern scientific vision produce less refuse than those with poor and dirty living habits. Based on the comprehensive survey conducted by it on the overall physicochemical composition of the refuse, NEERI has recommended scientific conversion of refuse into compost.

Accordingly, a compositing plant with a capacity of 200 tons per day of manure production has been installed at Calcutta. All other big and medium size Indian cities must follow similar practice to keep their cities clean.

Radioactive substances produced during fission and fusion enter into the life supporting environmental cycles. Nuclear fission is a process in which heavy nuclides are broken into lighter ones whereby energy of the order 19°K Cal is released.

On the other hand, a process in which lighter nuclides combine together to form a heavy nuclide is known as nuclear fusion. The high temperature of sup is maintained through nuclear fusion whereby, hydrogen atom is converted into helium.

All elements released into environment do not enter into the food chain, rather many of them go into marine ecosystem and many of them are so diluted by nutrient rich terrestrial ecosystem that they hardly remain harmful.

Examples of such harmless radioactive elements are Manganese-54, Iron-59, Cobait60 and Zinc-65, those that are present in particulate or collodial from e.g., Parasedodymum-144, Zironium-9 and Rhodinm 106, which get transferred into the marine ecosystem in the highest amount.

In contrast, the soluble fission products such as Strontium-90, Cesium-137 are found in highest possible amount in land and plant. Further, Cesium-137 gets accumlated only in certain plants like mushrooms and lichens and hence does not pose as great a hazard as Strontium-90.

The greatest hazard of Sr-90 lies in its accumulation in the human system through various plants and foods like, milk, grass vegetables etc., and creates a negative balance of calcium present in the human bone.

FOOD POLLUTION

Food ranks third in position, after air and water, for the maintenance of the process of life on this earth. Strength to tight against diseases depends on the intake and purity of the vital constituents of food, such as proteins, fast, carbohydrates, vitamins and minerals.

Removal or denudation, in part or in whole of any of the nutrients from food articles, or addition of adulterants adversely affect our health. The strength of a nation is by and large dependent on the health of its population.

The proverb "Health is Wealth" and "sound mind in sound body" are well known to us. Therefore, maintenance of the purity of food and preventing adulteration by harmful ingredients are direct contributions to the national health development programme.

Food adulteration act has long been enacted (1954) by the Government of India, under which food adulteration is an offensive criminal act and is liable for deterrent punishment.

Adulteration, Pollution and Contamination of Food

International mixing of foreign substances with food materials, for undue economic gain is known as adulteration. The entry into the food chain, of undesirable substances through environmental agencies i.e., air, water and soil is termed as food pollution.

The production of toxins in food through microbial agencies is known as food poisoning. The degradation in quality of food materials under the microbial influence to such an extent so as to produce diseases on consumption is termed as contamination of food.

The pollutants may be of chemical and biological types. Both types of pollutants enter into the food materials either through environmental routes or by adulteration. Some of the pollutants have been reported to be naturally present in harmful concentration in some of the food materials.

The potential chemical ingredients present in food, as pollutants are flourides, nitrate, pesticides, radioactive isotopes on many trace element, e.g , selenium, lead, mercury, chlomium etc. Naturally present organic pollutants in some of the food are—solanine, gossypol, oxalic acid and erucic acid.

Commonly used food adulterants are argemone oil, tricresyl phosphate and white oil for mustard oil, vanaspat (hydrogenated oils) for butter, and chromium or lead for turmeric (haldi).

Toxins are produced in food by clostridium batulinum, cl. perfringens staphylococcus aureus, salm-onella and bucillus. The detail for the above mentioned, pollutants, adulterants, food addatives and bacteriological toxins are briefly presented below.

CHEMICAL POLLUTANTS

Flouride

Tea, spice, betal gram, french beans and fishes are rich sources of flouride. Fruits and vegetables are generally low in flouride content. Among cereals, flouride content is minimum in rich preparations while it is maximum in chapati made of wheat which contains 0.8.1.8 ppm of flouride.

A high intake of calcium, magnesium and vitamin C is known to play an important role in preventing chronic flouride intoxication. People, living in the regions where water contains more than 1.5 ppm of flouride, are advised to take citrus fruits which are rich in vitamin C. Besides, they are advised to avoid spicy food because it increases the flouride.

Lead

Presence of lead in human blood in more than 40 microgram per 100 grams blood can inhibit enzymes such as ALA which are essential for biosynthesis of haem (a basic unit of blood). Soft water, paints etc., also contribute lead to our food materials.

Cadmium

It is not required by the human body even in the trace amount and therefore is completely absent in an infant. The cadmium intoxication can considerably be prevented by increaing the intake of vitamin C. Cadmium burden in excess of 20 to 30 mg in the human body may create a negative calcium balance, leading to osteomalcia softening of the bones.

Multiple factures resulting from osteomalcia are responsible for the painful disease known as Itai-itai.

Tin

Oils, fats, ghee, beverages, jams, fruits, vegetables, biscuits confectionery, coffee, tea etc. are marketed and stored in consumer's houses in containers generally made of tin. Acidic food stuff are more prone to be contaminated by tin.

Canned moist food i e., butter, grape fruit juice etc. on opening starts dissolving tin. For example, canned tins apple juice when first opened may contain 20.3 mg of tin per kg of juice while after 2 days, the quantity may go upto 87 mg/kg and after 4 days, upto 364 mglkg.

The food stuff and beverages normally contain tin in amounts which are usually below 100 mg/kg and ai e generally well tolerated. The toxic does in food which are causes acute toxication is believed to around 2 to 3 grams/ day for weeks together.

ADULTERANTS

Argemone Oil

The argemone sects are black in colour and similar in size as mustard seeds any latter is therefore very been adulterated with the former, it has been reported that regular consumption for 1-3 months of an edible oil, contaminated with as low a concentration as 0.1 % of argemone oil can give rise to epidemic dropsy in human beings.

Dropsy is a disease, causing swelling of feet, thighs and face. If the consumption of adulterated oils is not discontinued, it may cause glaucoma (disease of the eye) leading to blindness or swelling of liver which may be fatal.

Tricrecyl Phosphate (TCP)

TCP is practically colourless, odorless, non-volatile and stable liquid, insoluble in water but is freely miscible in vegetables oils. It is used as adulterant with the mustard oil. Consumption of oils containing TCP for at most 20 days may cause flabbly paralysis. Like parathion (a phosphorous pesticide), it may also impair the nuero-muscular control of the body.

Vanaspati (Vegetable Oil)

The fundamental difference between ghee (butter) oil, vanaspati is that the former contains saturated fatty acids whereas the latter contains such in free state. Vanaspati unlike ghee, contains trans-acid which may cause adverse effects on the consumer's health on prolonged use.

Vanspati is not only cheaper than ghee, but also closely resembles the latter in appearance. Vanaspati is, therefore, commonly used for adultering ghee. Such adulteration can very easily be detected by the following method even without going through any complicated chemical tests as such:

> Ghee is taken in a test tube and is melted in a burner or spirit lamp. One spoon full of sugar and 10 cc of concentrated hydrochloric acid are added into it. The mixture is then warmed for two minutes. Development of red colour confirms the adulteration of the ghee with vanaspati.

MICROBIAL TOXINS

Botulinus Poisoning

Such poisoning is caused by colostridium botulinum (an anaerobic spore forming organism) in improperly sterilized preserved foods which have a near neutral or slightly alkaline condition.

The poison caused by cl. botulinum is most potent because ingestion of even minute quantities would result in a severe syndrome that may be fatal within 2 to 10 days.

Staphylo-coccus Poisoning

Stapylo-cocci bacteria grow in cream, puddings, sauce and other starchy preparations. Besides these, milk, egg and meat are also prone to their attack, especially during summer, warm foods, kept for long, at room temperature, in summer months are open invitations for such microbes to make the food poisonous.

In contrast to salmonellosis or butulism the attack does not last more than one day and usually wanes within 6 hours or less.

NATURAL ORGANIC TOXCANTS PRESENT IN SOME OF THE FOODS

Solanine

It is a common name used collectively for a group of tozic glycoalkaloids found in a number of plants. Potatoes damaged or black-spotted potatoes may contain upto 400-500 ppm of this toxicant. Consumption of such potatoes may cause abdominal pain and diarrhoea.

Gosypoll

It is a toxicant made up the phenolic compounds and may be found in free state as well in combination with protein, in cotton seed and its flour, but never in cotton seed oil. It is toxic to man, pig sheep, chicken and fish, but not to the ruminants (cows and buffaloes).

The maximum allowable concentration of this toxicant is 1.1 per cent in cotton seed flour used for making biscuits.

Oxalic Acid

It is contained in chana dal (180 mg) piyas (407 mg), amla (296 mg) and khesari oil (lathyrus 122 mg) per 100g. *Consumption* of heavy dose of the food produces *acid cramps, haemohiliae* and even death. But such poisoning is generally rare.

However, it is advisable to restrict the intake of food containing more than 0.5 per cent of the acid. Incidence of formation of stones in kidney is high among the poor population of Rajasthan, presumably because of their habitual heavy consumption of the above mentioned food items.

Erucic Acid

It is a long chain unsaturated fatty-acid normally present in mustard oil mixed with rape-seed oil in concentration exceeding to about 45 per cent. Habitual use of such mixed oil has been found to cause myocardial fibrosis (reduction in eye sight) in human being.

NOISE POLLUTION

Noise pollution is often misinterpreted by most of the people as sound pollution. As a matter of fact, sound is a pure tone, harmonically related, having fixed frequencies and amplitudes, occurring at regular intervals and produces meaningful communication and pleasure in hearing.

Musical lectures given by teachers, leaders and the preaching by saints are the examples of sound. On the other hand, 'unwanted sound' is termed as noise, having a complex mixture of a number of pure tones of various frequencies and amplitudes.

These sound waves fluctuate and repeat themselves in highly haphazard manner. Noise pollution has become a part of industrial life in modern times *Urban people* are at constant exposure to such pollution contributed by disorderedly maintained vehicles Rural population are *increasingly* being exposed to noise pollution by '*loudspeakers*', ail amplifier of music especially during social functions.

The people working in *non-industrial* establishments are also vulnerable to noise pollution, significantly contributed by ill-maintained type-writers and electronic instruments.

Units of Noise Pullution Measurement

The unit for the expression of the Noise pollution intensity is decibel (db) *i.e.* the logarithmic ratio of noise intensity (I_0) to reference level (1). The minimum intensity of sound audible by man is termed as reference level (1), approximately equal to the sound pressure of 0.0002 microbars (dynes per cm^2 of an energy of about 10^{-16} watts). *Mathematically* the db is expressed as

$$db=10 \log_{10} \frac{10}{1}$$

Accordingly, 10, 20 and 100 db represent 10, 10^2 and 10^{10} times the threshold intensity respectively. The threshold intensity of sound required for human hearing is zero db, the rattling of leaves produces 10 db, and sounds generated during broadcasting is 20 db.

Sound level of 30 db is generated in bed room at night pillow talk. 40 db in library and 60 db in normal conver!ional speeches. The play of the radio at average volume induces a sound level of 70 db, light traffic produces noise at 74 db and 80 db of noise is produced by a typical factory.

Running train generates a noise of 90 db, orchestra of 100 db, rock band of 110 db while an aircraft take off produces noise of 120 db. Threshold of pain is of 140 db.

Measurement of Noise Pollution

The instruments needed for the measurement of noise pollution are many and varied, depending upon the types of experiments to be done. For examples, *sound level meter*, *noise dosimeter* and *frequency* analyser are the instruments, commonly used for measuring the total intensity of noise, the actual noise that the worker receives during working hours and the frequency bands respectively. Brief description of each of the instruments is given below

Sound Level Meter

A sound level meter consists of a *microphone, amplifier, rectifier* and a *display meter*. The sound pressure level in db is directly read out from this instrument.

Noise Dosimeter

The different noise levels, existing close to or away from the machines are measured directly by this instrument. The total levels of noise that wearers receive during their working periods are indicated whereas the total intensity of sound produced in the campus are shown by the sound level meter.

Frequency Analyser

The frequency-cum noise level distribution is measured by such type of instrument which helps to devise any control measures to reduce the potentially effective part of the noise level.

Acceptable Noise Standard

Sound is essential for life. Organisms are at constant exposure to some other forms of energy without which the brains may not function. At the same time the adsge. "Too much of every thing is bad" is also applicable in this case.

The unwanted sound', what is called as noise, is not at all ecologically desirable at any level. However, the permissible noise levels in db is dependent upon the time of exposure, 80 db of noise level for 8 hours days, 90 db of noise for 4 hours per day, 95 db of noise for 2 hours per day, 100 db of noise for 1 hour/day 105 db of noise for 1/2 hour per day.

110 ab of noise for 1/4 hour (15 minute)/day and 115 db of noise for 1/8 hour per day have been permitted by almost all the countries of the world. Exposures to continuous or intermittent noise level higher than 115 db should not be permitted.

However, 90 db of noise level for 8 hours/day of exposure has been granted in case of mining occupations for workers in India. The maximum desirable limit for noise levels in residential area and recreational activities centres around the habitations has been fixed as 40 db in night, 45 db in evening and 50 db during day time.

Effect of Noise Pollution on Human Health

Noise level above 80 db for more than 8 hours a day has been found to increase tension in a human being of normal mental make-up. Noise level above 80 db during intermittent exposure produces a change in breathing pattern.

Continued exposure to high levels of noise results in annoyance, fatigue, shifting of thresold limit of hearing and the worst may be the total loss of hearing. The bad effects of noise pollution have been seen to interrupt the process of *blood circulation*, to increase breathing amplitude to ;bring changes in size of the pupils of the eye and in the rate of saliva and gastric secretions.

Noise pollution above 120 db produced by aircraft and turbine have been shown to induce many adverse *biochemical* changes in a human being for 3.88 hourse of exposure per day for a period of one to 6 months.

The concentration of *cholesterol* in blood becomes elevated alongwith increased white cell counts in people living around close *vicinities* of busy *aerodromes*.

Industrial workers, exposed to noise level of more than 100 db, reveal the tendency of reduced *intra-ventricular* conductivity besides hypertension.

Remedies of Noise Pollution

(i) Decibel meters should be installed along highways and in places of public gatherings such as picture halls, libraries, offices and hostels. An industrial establishment must have such facilities in order to have a check on the intensity of noise pollution, being produced througout the working period.

(ii) Greenbelt vegetation and open space, at regular intervals, within the densely populated areas of big towns, must be developed. A dense evergreen hedge can reduce the noise by 10 db during garbage collection in the town. *Border planting* along highways can very effectively control the noise pollution if plantings be done in lower side with respect to the source of noise production and be on high side towards the listener.

(iii) A strip of 20 feet wide plantation inside the compound wall, effectively protects the houses from sound pollution, contributed by *vehicular traffic*.

(iv) Technological modification is suggested to be achieved so that a sound associated with minimum possible frequency be produced from the *machines*.

Chapter 2

FREE FLOATING ORGANISMS

The word '*plankton*' was coined by Viktor Henson in 1887. Broadly defined it includes all living organisms both plant and animal that *drift* about passively in water. Many of these organism have the ability to Swim, but because they are small in size *generally*, progression through the water is relatively slow and their distribution is often *patchy* and determined by the movement of ocean currents and tides,

Many planktonic organisms are microscopically small. but there is great variation in size. The pulsating bell of some of the larger *jellyfish* may be 2 m across with tentacles up to 20 m; these form part of the *megaplankton*.

A more 'biological' classification divides plankton into *phytoplankton* (*drifting plants*) and *zooplankton* (*drifting animals*). *Holoplankton* and meroplankton respectively are terms used to describe organisms that spend either their whole lives or only part of their life histories (usually the larval stages) drifting passively in the water. *Planktonic* larval stages provide a means of dispersal before the adults settle down.

NETS

How is this unevenly distributed plankton caught ? The simplest kind of plankton net is conical in shape, with a collecting vessel at the far end. As the net is pulled

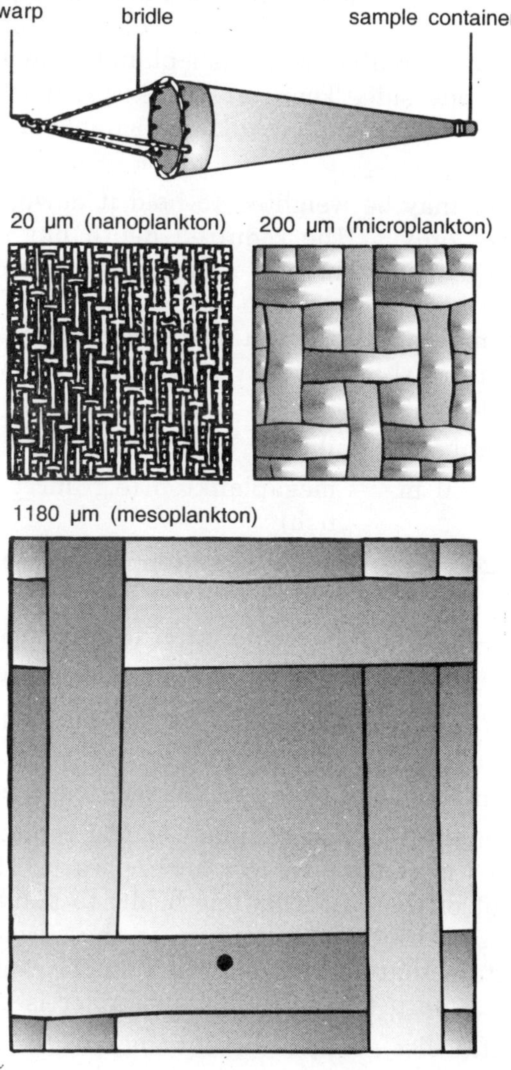

Figure 2.1: Conical plankton net (top). Types of monofilament polyester bolting cloth showing relative range of mesh sizes.

slowly, so the water passing through is filtered and any plankton caught on the meshes is *swept* into the collecting vessel. The *mesh* size will determine if the catch is largely macro or *microplankton*. Pores soon become *clogged* in very fine mesh netting and little water will be filtered because the bow wave created will push plankton in the path of the net to one side. There are other problems with this simple kind of net including taking a sample at a *particular depth*.

The net may be weighted to hold it down, but it may sam,,le not only at the required depth but also on the way down and up. There is no easy way of telling which specimens were collected at different depths when this mixed sample is being sorted in the laboratory. One means of overcoming this problem is by using a *Nansen-type* closing net.

Size range of planktonic organisms

0.2—2.0 m	=	megaplankton (e.g. large jellyfish, fish)
2—20 cm	=	macroplankton (e.g. some shrimps, fish)
0.2—20 mm	=	mesoplankton (e.g. copepods)
20—200 μm	=	microplankton (e.g. protozoa)
2—20 μm	=	picoplankton (e.g. autotophic flagellates)
<2μm	=	picoplankton (e.g. bacteria)

When the net has sampled at the required depth a '*messenger*' is sent down the towing warp. This changes the weight of the pull from one bridle to another and the net is closed prior 'to hauling up to the surface, In order to be certain that the net is closed on the way down and on the way up a *Clarke-Bumpus* sampler is used.

A metal disc sealing the entrance to the net may be rotated by sending a messenger; once the sample has been taken a further messenger causes the entrance to be closed by rotation of the disc. These *flap-valves* are restricted to nets less than 50 cm diameter for *mechanical reasons*.

Acoustic command signals are now often used instead of brass '*messengers*'. The amount of water that flows through the net is difficult to estimate, especially once the meshes of the net become clogged and a bow wave is set up. However, by fitting a flow meter at the cntrance to the net and by relating the size of the opening of the net and the reading on the flow meter, essimates of the volume of water sampled may be made by *multyplying* the mouth area of the net by the distance towed.

This is essential if any *quantitative* estimate is to be made of the amount of plankton in the sea. Non of these refinements is necessary if only a *quantitative* sample is required, one disadvantage of all these nets is that they may be towed at a few *knots* only, otherwise the water pressure brusts them.

In order to speed up sampling, Sir Alister Hardy devised a continuous plankton sampler which could be used at speed and towed behined *commercial* ships on their normal routes. It has a strong metal outer case and its streamline shape cuts down water pressure. A criss-cross pattern of cruises built up a good record of plankton distribution, especially in the *Atlantic* and in the *North Sea*.

A forerunner of the Hardy sampler was the 'plankton indicator' used by herring drifter *skippers* to locate waters rich in the copepod *Calanus* on which the *herring* fed. High speed sampling of plankton is possible using the Gulf III net made of corrosion resistant *cupro-nickel* alloy (hence the *abbreviation* TTN meaning '*tin tow-net*'.

The rigid outer casing allows the inner net of metal gauze to be better supported and take the strain imposed at high speed. A *nonporous mouth reduction cone* increases filtration efficiency by drawing a column of water wider than the reduced mouth, through the *sampler* Much emphasis is being placed currently on remote sensing.

Infrared photography from satellites, coupled with computer enhancement of images, has enabled the distribution to be plotted of phytoplankton blooms often associated with frontal systems between water masses.

Plankton is not confined to the open sea. It occurs in shallow water and even surf. A two-iered plankion sampler towed by hand in the shallows at the water's edge collects *plankton* washed close inshore. Static plankton nets may be used to collect organisms emerging from the sediment and performing vertical migrations into the water column. Pumping water through filters may be used also to sample plankton in the top room.

Planktonic Organism and Larval Stages

Bacteria

These are certainly to be found, but not usually in very large numbers in the water column because of the paucity of solid objects on which to settle. In a few localities such as in areas of upwelling off the coast of north-west Africa or the hydrothermal vents associated with the spreading of plates farming the earth's surface, high bacterial counts have veen recorded in some cases over a billion per litre.

The *Galapagos rift* in the Eastern Pacific is a result of such *spreading*. A combination of the minerals in solution and the suspended bacteria give a milky blue tinge and shimmer to the hot water ejected from the vents in the rift.

Chemosynthetic bacteria in the vent water utilize hydrogen *sulphide* as a source as energy. This contrasts with the more usual dependence of food chains on *solar energy*.

Phytoplankton

This comprises all those-tiny single-celled organisms containing chlorophyll and other pigments which like land plants, are able to utilize carbon dioxide, nutrient salts and the energy of sunlight to produce carbohydrates, proteins and fats.

Most photosynthetic activity occurs in the 15 m of the sea but some occurs even at 100 m in clear oceanic waters. There are two main groups of phykoplankton : diatoms and flagellates.

Diatoms

These live mainly in the surface waters and their silica skeleton is often elaborately *ornamented* which may have the effect of slowing down rates of sinking. Some contain a drop of oil which improves buoyancy. Reproduction is by simple asexual cell division. The cell contents of two diatoms emerge from their silica 'shells' after the nucleus has undergone a reduction division *halving* the number of chromosomes.

A nucleus from each partner fuses and produces a nucleus with the full complement of chromosomes, The protoplasm of this newly formed zygote swells and secretes a silica outer skeleton of normal size. Each diatom '*shell*' is made up of two halves and when the cell divides each half makes another to complete the outer *skeleton.*

After several generations certain of the diatom '*halves*' become smaller- and smaller but before becoming *infinitesimally* small, the process is halted by resorting to *sexual reproduction.*

Flagellates

These are more mobile and use their whip-like, flagella to provide propulsion through the water. It was a thought that all life on land and in the sea depended ultimately on

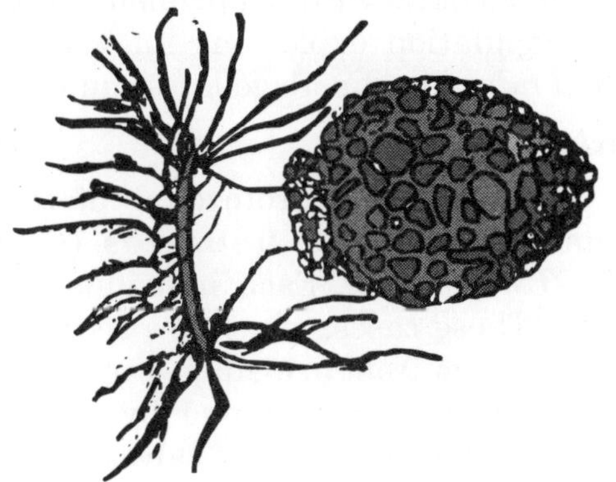

Figure 2.2: A tintinnid protozoan.

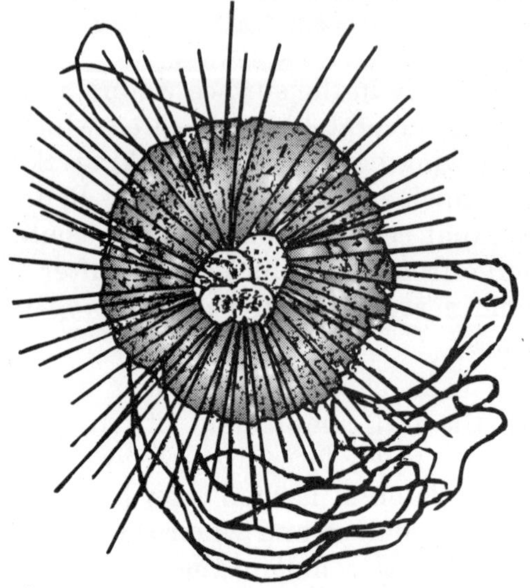

Figure 2.3: A living foraminiferan.

the ability of plants to '*fix*' sunlight energy by means of the a process of photosynthesis and that these diatoms and flagellates existing in countless millions in the surface waters were the primary producers, providing energy for the entire marine food web.

However, chemosynthesis by autotrophic bacteria may y play a role in powering some local deep sea communities. *Dinoflagellates* with red pigments occasionally undergo w vast population explosions known as blooms which produce '*red tide*'. The dinoflagellates can be removed.

Micro-zooplankton

In oceanic waters ciliated protozoans of the micro-zooplankton consist largely of *tintinnids* (*subphylum Giliophora order Tintinnoida*). Each individual is housed in a case of tectin. These cases, which are secreted by the animals and vary in form from urn-shaped to tubular, may be plain or more often strengthened by *agglutinated* material, such as sand grains, *diatom* valves or coccoliths.

Tintinnids range in size from 20 to 640 microns and

most species live in warm temperate or *tropical regions*. At times they become very abundant in coastal waters and the surface layer of the open ocean. Their main food, which consists of organic *particles*, *bacteria* and the *smallest species* of the *phytoplankton*, is wafted to them in currents created by their apical complex of *cilia*.

'The remarkable evolutionary versatility of the foraminifera has enabled them to *colonize* virtually every kind of marine environment with the result that fossils *foraminifera* are more nearly universally present in the marine sediments of the last 100 million years than any other group of *fossils*'.

Living foraminifera are known from more than 1,200 species, of which thirty or so form part of the plankton of *oceanic waters* : the rest live at all depths on the ocean floor and will be considered under benthic animals. All species secrete a chitinoid test around the main body of *protoplasm*.

The test is pierced by one or more holes through which the photoplasm flows out to form the filamentous pseudopodia. In the planktonic species and others the test is *strengthened* by *calcareous* material. Many of the benthic species have *agglutinated* tests.

Though *foraminifera* range widely in size from 20 microns to 190 millimetres, the mean size of the common species is about 500 microns.

The *planktonic* species live in tests formed of a cluster of interconnecting chambers, which in many species are spirally called and look like *miniature* snail shells. Beside the main test aperture, the walls of the *chambers* are *punctured* by many small holes through which emerge the *pseudopodia*.

When fully extended these *cytoplasmic filaments* not only radiate in all directions, but flow into one another to form a network. It seems that the main food of the surface-dwelling forms consists of *phytoplankton* and small *ciliates*, which are trapped by the *pseudopodia* and digested on the spot.

Planktonic foraminifera are most abundant in the epipelagic zone, where most, if not all, species are thought to spend their early life. The *epipelagic* species *generally* have spiny tests and the spines not only support an outer, *frothy cytoplasm* but *presumably* help to keep the rate of sinking to a low value.

The species without a spiny test have a greater depth range and some (e.g. *Globorotalia crassiformis, G. scitula* and *Hatigerinella digitata)* are evidently mesopelagic or bathypelagic forms. Most species live in tropical and *subtropical* waters, which overlie most of the vast expanses of calcareous sediments that are rich in the shells of planktonic *foraminifera* (*globigerina ooze*) and cover about 45 per cent of the deep-sea floor.

Living *Radiolaria,* which are known from about 4,400 species, are exclusively marine and with a few exceptions planktonic. The very numerous fossil species go back to

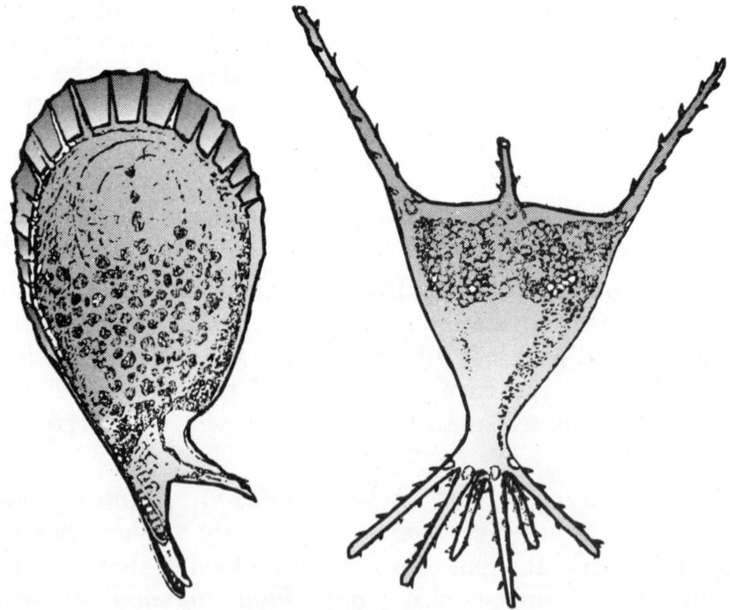

Figure 2.4: Two deep-sea radiolarians.

Precambrian times. In nearly all living species the endoplasm is housed in a central capsule: the *ectoplasm* contains, or is contained by, a *gelatinous* layer.

As in foraminiferans, long and fine pseudopodia radiate outwards and finally anastomose to form an elaborate food *trapping net*. Some species are *luminescent*. All but a few species *secrete* some kind of skeleton formed of spicules that may be isolated but are much more frequently joined in diverse and beautiful forms, many of concentric and radiating patterns. *D'Arcy Thompson* was reminded of *snowflakes*.

'The little skeletons remind us of snow crystals (themselves almost endless in their diversity), rather than a collection of animals constructed in accordance with functional needs and distributed in accordance with their fitness for particular situations.

The unity in diversity of radiolarian skeletons that so intrigued Sir D'Arcy was rediscovered by *Buckminster Fuller*, who was charmed to find geodesic structuring in their skeletons, and in those of diatoms. Indeed, the skeletons of numerous radiolarians look very like the geodesic domes designed by ckminster Fuller and others.

As he wrote: '*Employing* the fundamental principles of nature's structuring, the average of all the geodesic domes to date enclose space at approximately one per cent of the weight of materials required by conventional architectural and engineering structures of equivalent volume designed for *equivalent* load stresses and clear span environment controls.

Upheld by the sea, *radiolarians* face no such stresses, but they presumably need to be supported by skeletal designs that make the best possible use of rare supplies of building material. *Curiously* enough, the skeletons of the *Acantharia* (sometimes classified as a separate order), which use rare *strontium* in the form of sulphate, look more like snow-crystals than *geodesicdomes*.

But the radiating configuration of spicules is integral to

their organization: the spicules support a well-formed *gelatinous* layer that surrounds the central protoplasmic body. In most acanthárians contractile threads (myonemes) are attached between main *spicules* and the surface of the *gelatinous* layer, the entire system forming an '*appareil hydrostatique*'.

The skeleton of the three other main groups of *radiolarians* is made of silica. The *spicules* are isolated in certain species but in most are united, often in geodesic array, to form *concentric lattices* or shells of\ *extraordinary delicacy* and *diversity*.

A central *capsule* is always present. The *ectoplasm*, part of which may form a frothy layer that may help in *flotation* usually contains a well-developed *gelatinous* layer, The species of two orders. like the *acantharians*, are *predominantly epipelagic*. The former are very diverse and include many species with *concentric*, *latticed shells*.

The deep-sea species belong to a third order, the *Phaeodaria*, *notably* to the families *Challengeridae* and *Tuscaroridae*. Species of the first family, which have an oval or lenticular shell pitted with minute holes, live at *mesopelagic* and *bathypelagic* levels.

In species of *Challengeria* the size increases with their depth range. The tuscarorids, with a spherical, elliptical, fusiform or pyramidal shell, are also small forms with a wide depth range. Like the foraminifera, epipelagic radiolarians trap and digest microscopic plants and animals in their pseudopodial complex.

They are said to take flagellates, diatoms, tintinnids and even young stages of copepods. In species other than *phaeodarians* the protoplasm contains the resting stages of *dinoflagellates* known as *zooxanthellac*. The latter are *photosynthetic* and there is presumably some form of nutrients exchange between plant cells and their hosts. Radiolarians are also most diverse in subtropical and tropical regions.

At times they are very abundant in productive areas

but little is known of their overall success in oceanic waters, During a crossing of the western North *Atlantic* catches of *radiolarians* were *generally* less than those of *foraminiferans.*

The formation of the siliceous oozes rich in radio larian skeletons reminds one most of Rachel Carson's evocation of the '*long snowfall*' that forms the deep-sea sediments. 'When I think of the floor of the deep sea, the single, overwhelming fact that possesses my imagination is the accumulation of sediments.

I see always the steady, *unremitting*, downward drift of materials from above, flake upon flake, layer upon layer-a drift that has continued for hundreds of millions of years, that will go on as long as there are seas and continents.'

It seems odd that radiolarian ooze is almost entirely restricted to the deep-sea floor below the productive waters of the *North Pacific Equatorial Divergence.* There are also substantial deposits of radiolarian ooze on deep stretches of the ocean floor below the rich waters of the Peruvian Current. Why is so little *radiolarian ooze* produced in the other oceans ?

The Zooplankton

Some biological categories are more arbitrary than others, but if the micro-zooplankton is taken to include organisms that will pass through a net with a mesh size of 5'0 microns, the net, if large enough, would retain all the adult forms of zooplankton, which range in size from 300 micron copepods to the jelly fish *Cyanea capillata,* with an umbrella that may reach a span of about 2 metres.

If we remember that the tentacles of a *Cyanea* may extend for 50 metres or more, the size spectrum of zooplankton covers more orders of magnitude than that of the nekton, with a range from a squid or fish of a few centimetres to a 35 metre blue whale.

In taxonomic terms the zooplankton virtually consisotsf the pelagic forms of *Cnidaria, Ctenophora combjellyfishes, Polychaeta, Mollusca*, small *crustaceans*, *Chaetognatha* (*arrowworms*) and *tunicates*. Young stages in the life history of these animals also form part of the *zooplankton*, but are considered elsewhere under early life histories.

The concept of plankton, as already stated, centres on organisms that drift in the *hydrosphere*. In the ocean the speed of water movements ranges widely from quick-moving surface currents, to the slow circulation of the depths, where velocities may be no more than a fraction of a centimetres per second.

Velocities of 10 to 50 cm/sec are very common, which are well above sustained swimming speeds of most forms of zooplankton. Though the zooplankton moves with the sea, and indeed, certain species make reliable indicators of water movements, this passive tendency is transcended by strategic activities whereby any species tends to maintain its population and centres of distribution in a moving ocean.

In his comprehensive review of propulsion in aquatic animals Lighthill shows clearly that the predominant and most successful form of locomotion is by undulatory movement. The smallest forms, such as the flagellates and ciliates, propel themselves by *undulation* of flagella or cilia. Fishes and whales undulate body and fins to make their way. Annelid and nematode worms throw their bodies into waves.

Though propelled by jet reaction, 'the second main alternative to undulatory propulsion in water', squids and certain octopods have fins that are fashioned for undulatory propulsion. In moving through water, small forms, whose relative motion is very slow, are faced almost entirely by the *viscous* resistance of their medium, which varies as the first power of their velocity: larger, faster-moving forms have mainly to contend with the inertial resistance, the resistance of the medium to displacements caused by a moving body, which is proportional to the square of the velocity The relative effect of these two forms of resistance is expressed by the Reynolds' Number R.

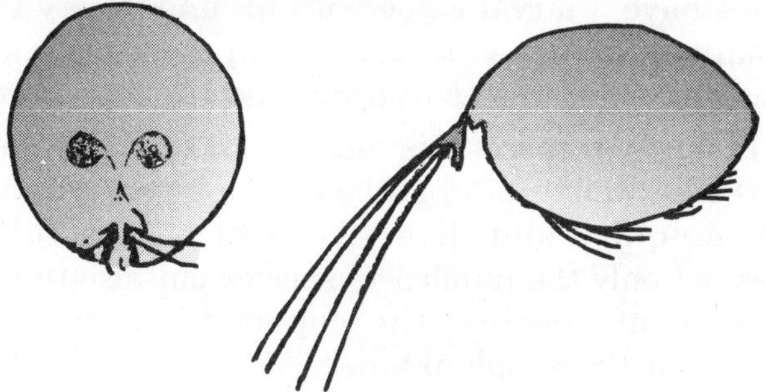

Figure 2.5: Ostracods.

$$R = \frac{LVP}{\mu}$$

where L is some dimension (usually length) defining the size of the body, V is the speed relative to the water is the kinematic viscosity, which is about 0.1 in water.

The undulatory mode of propulsion is thus effective over a wide range of R, which is one reason, as Lighthill says, why it is the principal mechanism in the motions of aquatic animals. Jet propulsion is appropriate only in animals large enough to yield relatively high values of *Reynolds' Number*. In this context it is interesting to survey the locomotion of planktonic organisms.

The smallest forms of phytoplankton (nanoplankton) are predominantly flagellates, while the rest are also flagellates or non-motile. The main groups of *micro-zooplankton* are either *ciliates* or without means of active movement (*foraminiferans* and *radiolarians*).

In the zooplankton, however. the small crustaceans and many of the gelatinous forms, which, as we shall see, are dominant groups, have other than undulatory means of locomotion. The *copepods*, meaning oar-footed, are well named, while the much larger cnidarians and the largest forms of tunicates (Thaliacea) have evolved *gelatinous* bodies that are shaped to muscular means of jet *propulsion*.

Even so, the cnidarians usually start their free.swimming life as small ciliated larvae (*planulae*) and some of the tunicates have a larval stage with an undulatory tail. The two dominant groups of undulators in the zooplankton are the smallest tunicates (*Larvacea*) and the arrow-worms.

If biological success is judged by a combination of biomass (the weight of organisms in a given unit of living space), ubiquity and diversity, which in a full sense involves not only the number of species but also the relative abundance of species, the copepod crustaceans are unrivalled in the zooplankton.

After the small crustaceans, we shall consider the gelatinous forms of zooplankton (*cnidarians*, *ctenophores* and *tunicates*), and the arrow-worms, which in varying respects meet some of the above criteria. The *polychaete* worms and *molluscs* are much more successful as *benthic forms*.

Small Crustaceans (Copepoda and Ostracoda)

Like their arthropod relatives, the insects, spiders and so forth, the Crustacea have segmented bodies and paired, jointed *appendages* that are covered by a skeleton made largely of *chitin* and *protein*. The body is divisible into *head*, *thorax* and *abdomen*.

In common withd most arthropods, the crustacean appendage is essentially a jointed system of hollow levers, that are moved by internal striated muscles. *Crustaceans* are readily distinguished from other *arthropods* in developing two pairs of *antennae* on the head.

Close behind the mouth is a pair of mandibles followed by two pairs of maxillae. *Copepods* have a dorsal carapace that extends over the head and one or two *thoracic segments*. The first pair of appendages on the thorax (*maxillipeds*) are succeeded by five pairs of *biramous* swimming legs in the free-living forms.

The abdomen is without appendages. There are about 6,000 species, most of which are marine. Individuals of the pelagic species (about 750) range from about 0.5 to 17 mm in length and live in virtually every part of warmer regions.

Much more often than not, copepods dominate the

zooplankton. For instance in standard net samples taken during a cruise between New York and Bermuda the volume of epipelagic copepods (measured by displacement) averaged 51 per cent of the total volume of zooplankton.

Copepods are also likely to be prominent in catches at mesopelagic and bathypelagic levels Most species are virtually transparent, but when their tissues become cloudy after preservation, they look like small grains of rice. Some of the deep-sea species are tinged in varying degrees with red pigments; a few are black. In some species there are luminescent glands in the skin such that the extrusion of their products causes flashes of light.

When copepods are cruising the five pairs of biramous thoracic limbs, which are flattened, paddle steadily backward in sequential strokes. If the animals are disturbed, these limbs beat strongly backward in rapid succession: at the same time the antennae are folded backward along the body.

The result is a relatively long dart, and for human eyes at least, such movements in transparent copepods are over in a virtually undetectable flash. The long leap of *Gaussia princeps,* a black, deep-sea copepod that teaches a length of just over 10 millimetres, is quite startling.

Copepods of the neustcn may even jump out of the sea. Wherever they are, these sudden bursts of speed must often enable copepods to evade their many predators in the zooplankton. Most of the pelagic copepods (calanoids) have long and well-developed first antennae, which may be suddenly flexed to jerk the animals forwards.

When copepods are feeding these antennae are extended outward, and no doubt help them to find their food. Species that feed mainly on pbytoplankton are generally the most abundant. The classic instance is *Calanus finmarchius,* which in north temperate and subarctic waters forms much of the food of large whale-bone whales and herring.

Herbivorous copepods filter their food through the long fine bristles of their maxillipeds but if the phytoplankton

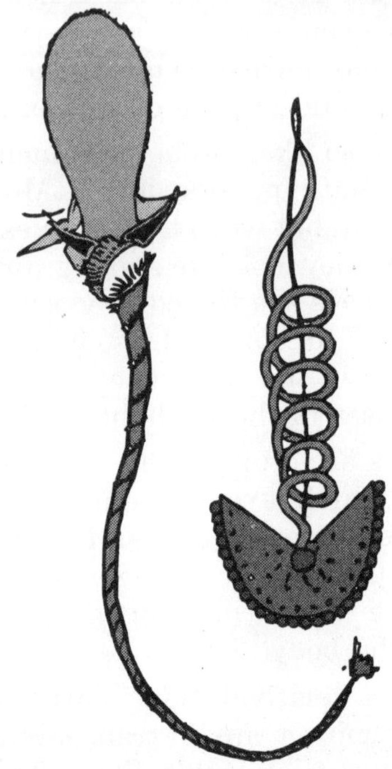

Figure 2.6: Right, a stinging-cell (nematocyst) of a cnidarian. Left, a sticky-cell (colloblast) of a comb jelly (ctenophore).

cells are large enough, they are seized one by one. There are also omnivorous and carnivorous species.

Members of all three tropbic groups exist together at surface and deep-sea levels of the ocean. In the tropical ocean the number of copepods in a cubic metre of water falls from about 15 in the surface waters to about 2 at lower mesopelagic levels (500-1.000 m).

Between 1,000 and 8.000 metres the number ranges from 0.42 to 0.05. The considerable success of some mesopelagic species is no doubt related to their daily vertical migration in search of better food supplies above the daytime levels of their populations.

Ostracods, which have much the same size range as copepods, are very differently made. The indistinctly segmented body is housed in a laterally compressed, bivalved *carapace* that may be closed by an adductor muscle.

There are two pairs of thoracic appendages but the antennae provide the main propulsive power. Most of the more than 2,000 species are marine and benthic in habit. Ostracods rival the copepods in ranging adaptability, but the planktonic species are not so abundant. At depths below 1,000 metres the numbers of copepods in a cubic metre are generally much more than ten times those of ostracods.

Some of the deep-sea species are luminescent and red or brown in colour. The largest form, *Gigantocypris,* which an almost globular carapace that attains a length of 1 cm, is a bright red *Gigantocypris* normally propels itself with a pair of fan-like antennae *protruding* through a slot in the carapace at the anterior of the animal. Balanced strokes propel the animal forward and movement of one or other antenna individually serves to rotate the animal round its vertical axis'.

Most planktonic forms are eyeless, but the eyes of *Gigantocypris look* like '...two huge forward facing, headlamp-like reflectors, which perhaps assist in its capture of the small fish, cbaetognaths, and copepods found in its stomach'.

If pelagic ostracods are generally carnivores or scavengers it is natural that they are less numerous than copepods. Compared to the herbivorous or omnivorous copepods their place in the food chain is less close to the primary source of food, the *phytoplankton.*

Like the copepods, some of the ostracods undertake daily vertical *migrations.* At one station in the eastern North Atlantic, species of *Conchoecia* especially *C. curia,* formed most of the food of the hatchet-fish, *Argyropelecus aculeatus.*

Gelatinous Forms

The gelatinous forms of zooplankton are cunningly fitted for a marine existence. Though usually jelly-like and fragile, the bodies of cnidarians, ctenophores and tunicates are supported and kept in shape by the sea.

In cnidarians the outer and inner tissues of the body wall are separated by a gelatinous layer, which is greatly developed and shaped as a matrix for the parachutes, bells and other parts of planktonic medusae, jellyfishes and siphonophores. Such organization is found also in the comb-jellyfishes.

In pelagic tunicates, gelatinous tissue is a matrix for the 'house' of larvaceans, the tunic of salps and doliolids and the tubular setting of pvrosomids. The water content of gelatinous tissue in cnidarians is 95 per cent or more. The matrix contains collagenous fibres that not only support it but give the jelly elasticity as well as a relatively high viscosity.

In tunicates the test, whatever its consistency, is made largely of a cellulose called tunicin. More generally, gelatinous tissue forms a hydrostatic skeleton that is very well fitted for propulsion by jet reaction. After each pulsation of the bells of jellyfishes and siphonophcres, which forces out a jet of water. the elasticity of the jelly and the return flow of water restore the bell to its resting shape in readiness for the next pulse.

Much the same is true of salps and doliolids in motion. When the gelatinous bracts of a siphonophore the detached from the colony they float upwards. They are positively buoyant, but how is this possible in tissue that seems to consist almost entirely of water and sea salts?

At Villefranche in 1957 E. J. Denton and T. I. Shaw found that gelatinous animals (like the large dinoflagellate *Noctiluca)* achieve their buoyancy by excluding some of the heavy ions of sea water, notably sulphate, from their body fluids and replacing these ions by lighter ions such as chloride. Denton

The results of the Villefranche experiments were quite clear. For five kinds of gelatinous marine animals, medusae, ctenophores, tunicates and heteropod and pteropod molluscs, the whole animals were usually either close to neutral buoyancy or positively buoyant.

This buoyancy could be quantitatively explained in terms of the salt composition of the animals' body fluids. These were always isomotic with seawater but contained less sulphate and more chloride, and also generally less magne sium and calcium but more sodium than seawater. In all of the animals studied the buoyancy in seawater mainly depended on excluding some sulphate from the body fluids and replacing this anion by chloride.

Presumably gelatinous tissue contains cells that exchange heavy for lighter ions. Whatever the mechanism. the energy needed for such differential ionic transport must be easily saved by the energy not needed, as it were, by buoyant gelatinous animals to keep their level in the ocean.

Indeed, the gelatinous nature of these planktonic animals must be seen as integral to their whole organization, which is relatively simple and easy to run. Most of the cellular elabora tion of a cnidarian resides not

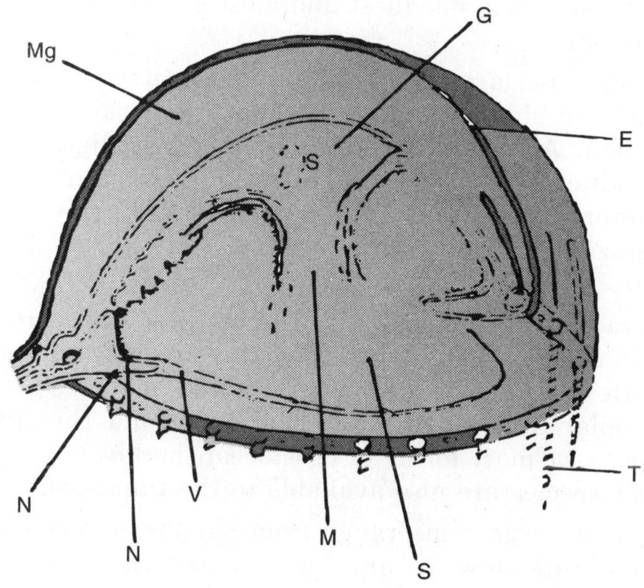

Figure 2.7: Structure of a trachyline medusa, based largely on Gonionemus.

so much in their means of locomotion but in their means of procuring and digesting food and in reproducing themselves.

Peter David's splendid photograph in *Deep Oceans* of a deep-sea jellyfish with a lantern-fish ensnared in its tentacles is a striking illustration of a simply organized animal availing itself of rich and long-lasting nourishment in a highly organized one.

The nervous system and muscles of the prey alone, which was just starting to be digested and is about as long as the bell diameter of the predator, are many times the biomass of the simple nerve net and muscle cells involved in the movement of the jelly fish and its tentacles.

One muc3 not, of course, forget the outlay of materials needed to build the millions of stinging cells in the tentacles (though nematocysts once formed need no maintenance, for they are dead motile units). Even more than their bottom-dwelling relatives, the cnidarians of the plankton are the simplest, deadliest and most effective predators in the ocean.

Tunicates are just as cunningly contrived. They are surely the simplest and most effective herbivores in the plankon. As in cnidarians, the feeding, digestive and reproductive systems rather than the nervous and locomotory systems, subsume most of the cellular organization of a tunicate.

By means of a house with a system of fine filters (Larvacea) or by means of a mucus filter net secreted by the endostyle (Thaliacea), tunicates are able to trap the smallest and often very abundant members of the phytoplankton. Such plant food is beyond the filtering capacity of most forms of crustacean herbivores, and the larger species are also available to the tunicates.

Living organisms range from simple to very complex forms, but how is one to express their degree of organization? Some measure may be got by estimating *cellular* quantities of deoxyribonucleic and (DNA), which contains the genetic blue print of all levels of organization in living forms.

The more complex the organism the greater is the amount of DNA. At extreme ends of the scale, a human cell contains over two million times as much DNA as that in a simple virus. *Sponges* have about ten times *as* much DNA as bacteria.

Cnidarians have about six times as much as sponges and in the present context, about twice as much as tunicates, which is remarkable considering that *tunicates* are related to the chordate ancestors of *vertebrate* animals.

Evidently, DNA content may be a measure of the degree of organization, which is usually but not always related to position in the *evolutionary* scale.

Simple organisms are easier to run than more complex forms of about the same size, and one way of expressing this difference is by measuring their rate of oxygen consumption. In the jellyfish *Aurelia* one such measurement is 0.0034 and in a tunicate (*Salpa*) 0.008, which are among the lowest figures for marine animals.

This is not altogether surprising. The jellyfish, for instance, essentially consists of simple layers of living tissue dispersed over a massive *gelatinous* skeleton that is dead apart from its content of amoeboid cells.

Moreover, the tissues of a jellyfish or a salp are so thin that they can obtain all their oxygen by simple diffusion : they have no need of respiratory organs. As Schmidt-Nielsen writes, 'Organisms that are supplied with oxygen by diffusion only, protozoans. flatworms etc. are mostly quite small, less than a mm or so, or have very low metabolic rates, as jellyfish do.

Although some jellyfish can be very large, they may contain less than 1% organic matter, the rest is water and salts. They have a very low average rate of oxygen consumption, and the actively metabolizing cells are located along the surfaces where diffusion distances are relatively short.'

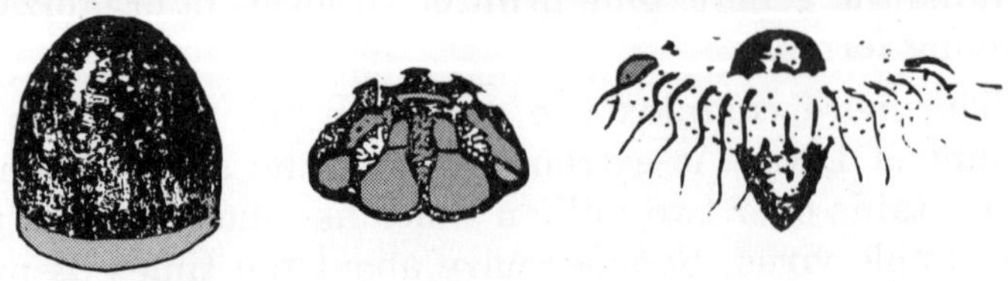

Figure 2.8: Three deep-sea medusae. From left to right, Crossota brunnea, Halicreas rotundatum and Atolla.

One can begin to see why planktonic tunicates and cnidarians are sometimes able to flourish in the deep blue deserts of the subtropical ocean. Thus, by a consideration of all the above attributes, the gelatinous forms of zooplankton are well fitted to their (predominant) place in the economy of oceanic life. But any attempt at a full appraisal must include some consideration of their size.

In one respect, as implied earlier. the size range of swimming bells conforms to the nature of the medium in that jet propulsion is effective only when principally resisted by the inertia of water. In the oceanic zooplankton the species with the widest size ranges are gelatinous and *jet-propelled,* which would still be true if one judges such range only on jet dimensions.

Even so, the aspect to be considered here is total size. The greater the size, and the cnidarians also have batteries of noxious deterrents, the less available they are to predators. Predatory neighbours in the plankton can only attack a jellyfish or a salp in piecemeal fashion.

To feast on jellyfish the predator must have a mouth as large as that of the ocean sunfish (*Mola*). Salps and medusae are sought by various kinds of large turtles. When the gelatinous forms of zooplankton evolved the means of making and shaping large amounts of their matrix, a wide range of size became possible, yet another biological advantage of their kind of organization.

Cnidaria

Cnidaria, from the Greek *cnides,* 'nettle', is an apt group name for hydrazoans, jellyfishes, sea-anemones, corals and so forth. There are about ten thousand living species, nearly all marine, and the key to their success and carnivorous way of life resides in the myriads of stinging capsules that form in the outer tissue layer.

While large forms, as we have seen, are able to trap, paralyse and feed on sizeable fishes, the overall success of the cnidarians would surely have been much less had they not evolved the right kind of nematocysts for securing the abundant supplies of proteins etc. below the armour of the smaller crustaceans.

In *Life in the Sea* by Gunnar Thorson there is a striking illustration of a small hydroid *(Hydractinia)* swallowing and entirely digesting a copepod, which is about as voluminous as the relaxed Lydroid, in the space of 26 hours. Copepods are strong and active for their size and must be poisoned and paralysed before they can escape.

Paralysis requires the injection of toxins below the cuticle of the prey. In a fish, on the other hand, the sensitive epidermis lies exposed over the scales and bones.

Nematocysts, the capsules of which range from ellipsoidal to globular in form and in size from 5 to 1.000 microns, are set on the tentacles and certain other parts of cnidarians. Each capsule contains a coiled-up thread with very thin walls.

When stimulated, and some nematocysts are triggered by a sensitive hair, the thread is forcibly and very rapidly everted. There are at least three main types of nematocysts : those that coil tightly and cling to the bristles on the surface of small crustaceans; those that fire a lassooing thread, stick or covered with barbed spines ; and those that pierce as they explode, so injecting the sting of their contents.

The threads of the last type are barbed to the tip and they rotate as they explode. 'Rotation of the tip must confer lateral and longitudinal stability on the advancing thread ; and it is striking to see bow straight are the tracks of discharges in distilled water, though the discharged tube is a most delicate structure.

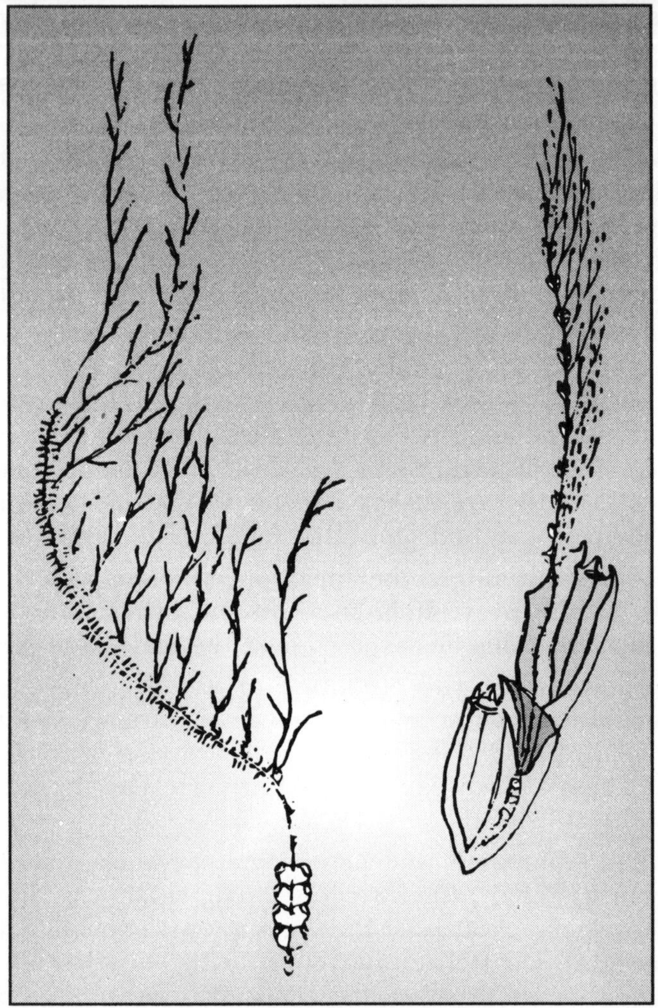

Figure 2.9: Representatives of the two main groups of siphonophores.

Rotation also imparts a cutter-like action to the barbs at the tip, as they pass to the barb series of the everted tube: this is probably unimportant in the lassooing type of capsule, but must determine the cutting and cuticle piercing powers of the injection type of stinging capsule'. There is much we do not know of the part played by mematocysts in the life of cnidarians, but we can at least begin to appreciate why they have held their place in the ocean from Cambrian times.

We have seen elsewhere that the outer tissue layer (ectoderm) of a cnidarian is separated from the inner layer (endoderm) by the gelatinous layer (mesogloea). The endoderm lines the body cavity and contains the digestive cells.

During development the mysteries of morphogenesis are such that these three body-wall layers are fashioned around the sea-filled body cavity into medusoid or polypoid forms. The adaptations of the gelatinous medusoid form for a planktonic existence should now be evident: polypoid forms, such as hydroids, sea-anemones and corals which have much less mesogloea, are primarily organized for a sedentary life and will be considered later with the other dwellers on the deep-sea floor.

But both forms are not only essentially similar in body plan but are also radially symmetrical, particularly in the placing of their tentacles. As Buchsbaum and Milne (1960) say, 'Their elegant symmetry is an effective design for snaring prey from any direction and passing it on to a centrally placed mouth.'

The main exceptions to the rule are the numerous *siphonophores* with *tentacles*, *digestive* and *reproductive* members dispersed along the trailing part of the organism. There are three classes of cnidarians. In the *Anthozoa*, which include octocorals, sea-anemones, stony corals and black corals, the life history centres on a *polypoid phase*. The *deepsea* forms are considered elsewhere.

In the Scyphozoa (*jelly fishes*) the medusoid form is predominant : the sessile polypoid stage produces young medusae, or develops directly to the adult condition, or is absent.

The groups of Hydrazoa that mainly concern us here are trachyline medusae, the siphonophores and chondrophores. Polypoid members of the important hydroid group live on the ocean floor at all levels. Shallow-sea species may have a medusoid phase (hydromedusae) in the life history. Hydroid medusae also occur in the deep sea.

For instance, during the 39th cruise of the Russian research vessel *Vityaz* in the Kurile-Kamchatka Trench area, such species as *Pandea rubra*, *Meator rubatra* (Bougainvillidae) and *Aequorea aequorea* (*Campanulinidae*) were taken.

Their relatives in the orders Trachylina and Siphonophora play a more prominent part in deep-sea life. *Hydrazoan medusae*, which range from disc-like to bell-like forms, differ from the jellyfishes, inter alia, in developing a velum.

The velum is a shelf-like inwards extension of the body wall around the rim of the medusa so as to restrict its opening to the sea. In the ectoderm of both the velum and the under surface (subumbrella) of the medusa is a circular band of striated muscle fibres, which act together to shape the swimming movements.

Quick contractions of these muscles purse the disc or bell to force water through the yielding and narrowing velar ring in a propellant jet. When the muscles relax, the elastic, viscous jelly and the inflow of water to the subumbrella space restore the medusa to its resting form in readiness for the next pulse.

Swimming movements may be desulatory or repeated rhythmically. The rate of pulsation changes with ambient influences such as the presence of food, the intensity of light and the temperature.

Size for size, bell-shaped medusae must be faster swimmers than disc-like types, for during pulsations a bell will approach a streamlined form and produce a narrow, relatively powerful, velar jet.

The form and structure of the sea-wasps (*Cubomedusae*), which are easily the fastest of the jellyfishes, support this contention). The bell-like body, which is disc aped in most jellyfishes, has four flattened sides and the margin of the bell is tucked into form a false velum (velarium).

Pulsations may be as much as 150 per minute and speeds of 1 knot or more have been estimated, making the fastest sea-wasps *respectable* members of the nekton. Thus, a bell-like form producing a concentrated jet and, of course, a well-developed neuromuscular system would seem to make for speed in *medusae* and jellyfishes. Close *comparison* of the fishing techniques of bell-like and disc-like species should be interesting.

Some disc-like forms (e.g. *Cyanea)* fish as their tentacles stream outwards during a slow parachute descent. *Hydroid leptomedusae* tend to be disc-shaped and have balancing organs (*statocysts*) round the margin of the disc. The other type (*Anthomedusae*) are bell-shaped with light-sensitive organs (*ocelli*).

Others troll long tentacles as they pulse in a *drifting sea.* Do the bell types, which presumably rely more on their own movements to carry out their fishing sweeps, catch much of their prey between pulsations, when their tentacles have chance to spread ?

Most of the trachyline medusae are high-seas forms with headquarters in the warmer oceanic regions at depths between the surface and bathypelagic levels. Bell widths vary from a few millimetres to 10 centimetres.

The species of one group (Trachymedusae) have a plain bell margin and include *Aglantha digitale, Halicreas minimum* and *Crossota brunnea,* all virtually cosmopolitan forms. The first, with a transparent bell, sometimes with rosy tints, is most abundant inthe epipelagic zone the populations of the other two, which have bowl-shaped bells, are centred at mesopelagic levels.

In *Hallcreas* the radial canals and tentacles are tinged with bright red pigments while the bell of *Crossota* is darkly coloured. Species of the other group (Narcomedusae have firm glassy bells with margins divided into lobes by the bases of the tentacles.

A large round leads to a capacious stomach. Forms that

occur at epipelagic to deep-sea levels include representatives of the gener *Aegina, Aeginura* and *Solmissus.* In the Adriatic Sea the daily vertical migrations of *Solmissus albaescens* have been assessed by over 100 net samples taken between 1.000 metres and the surface.

By day the populations are centred between depths of 400 and 600 metres, and during the afternoon individuale start to rise to higher levels. By night if the moon is full the medusae says below a depth of 10U metres but on dark nights swim nearer the surface.

This species, which has a well-developed velum with strong circular muscles, was estimated to ascend at a rate of 50 metres per hour, Copepods form the maid food, and the investigator thought that the medusae may have been following their prey.

Trachyline medusae which range from a few millimetres to 100 mm in width, are smaller than jellyfishes, but they evidently compensate for their limited prey-snaring radii by relatively high population densities, at least in highly productive waters.

In the North Pacific (Kurile-Kamchatka Trench), for instance, maximum numbers of *Aglantha digirale* were over 1,000 : lower down in the less food-rich mesopelagic gone the comparable figure for *Crossota brunnea* was 50.

The Siphonophora are pelagic hydrazoans formed of an inter-connected complex of medusoid and polypoid members that serve separate functions and are integrated as a colonial organism.

The polypoid members are modified as: digestive zooids (gastrozooids), with a mouth and usually a single tentacle; protective and prey-catching zooids (dactylozooids), slender, mouthless and armed with a tentacle full of nematocysts; and reproductive zooids (gonozooids), producers of the medusoid gonophores that form either eggs or sperm but never develop into complete medusae and are seldom set free.

Beside their modification as gonophores, medusoid

members form swimming bells (nectocalyces), with velum, four radial canals and ring canal, and bracts-gelatinous, leaf-like appendages that have buoyant properties. It is as though the separate medusoid and polypoid phases of the hydroids have been fused into one polymorphic organism.

The 150 or so species of siphonophores are predominantly inhabitants of the oceanic province from epipelagic to deep-sea levels. In two groups (Cyston:ctae and Physonectae) the colony is topped by a gas-filled pneumatophore or float and they are mainly distinguished by the presence or absence of swimming bells: physonects have one or more swimming bells below the pneumatophore, while cystonects have none. Members of the third group (Calycophorae) have one or more swimming bells at the head of the colony but are without a pneumatophore.

Cystonects include the Portuguese man of-war, *Physalia physalis, a* prominent member of the neuston and species of *Rhizophysa. Physalia,* which lives on the warm ocean but may drift into temperate waters, has a sail-topped float below which are tidily attached the tentacular, digestive and reproductive members.

In *Rhizophysa* these members are attached at intervals to a long stem that trails below a sizeable pneumatophore. Nearly all siphonophores are physonects or calcycophorans and most species resemble *Rhizophysa* in that the tentacular, digestive and reproductive members are hung on a long stem below the head of the colony.

More precisely, the gelatinous bracts cover these members to form separate clusters (cormidia) that are repeated regularly along the stem which is very much alive, sspecially in its remarkable powers of extension and contraction. Even in species with a float, much of the buoyancy of the colony resides in the bracts.

Physonects, such as *Nanomia* and *1lalistemma,* draw their line of trailing members through the sea by pulsing a relatively small number (15) of swimming bells. In more complex forms (e.g. *Agalma)* the bank of bells consists of more numerous units.

Certain forms (e.g. *Physophora* and *Discolabe)* have the nutritive and reproductive zooids clustered radially about a short stem below the pneumaatophore and swimming bells. *Nanomia* has been well studied.

Underwater films from submersibles show it capable of lively aquabatics. Evidently the activities of the various zooids are coordinated by the nervous system. Swimming bells pulse in unison to move the colony along a given line but are equally capable of asymmetrical action to change course suddenly.

Contractions of the tentacular members lead to lively activity of the neighbouring digestive members, and in turn digestive products are pumped to other parts of the colonial organism, surely one of the most fascinating forms of animal life.

Most species of siphonophores belong to the Calycophorae, some of which are the most widespread and abundant in the order. When mature the cormidia of most calycophorans break loose to become free-living sexual forms, and they are less prevalent than the asexual, colonial stages.

This is to be expected !f the production by gonophores of sexual cells, and their eventual union to form planula larvae, is seasonal. Also tae eudoxids are sometimes concentrated at deeper levels than their parent colonies. Species range in size and activity from small darting diph) ids to large and sluggish prayids.

The latter forms have two or more large swimming bells that are highly gelatinous and head a long line of trailing members extending for several metres. As meant by their name, diphyids have two swimming bells, which are usually similar and one is just before the other.

One species. *Chelophyes appendiculara,* may well be the commonest and most widely distributed species of siphonophore. Most siphonophores live in the warmer waters of the open ocean, where numerous species are widely distributed. Species of *Muggiaea* and *Nanomia* seem to be more neritic than oceanic.

The well-known species *Dimophyes arctica is* a coldwater species that lives at epipelagic levels in arctic and antarctic areas but is centred in cold, deep waters in warmer intermediate regions. It is one instance of equatorial submergence. In depth the populations of many species are centred in the epipelagic zone and come extend to mesopelagic levels.

Certain species are predominantly mesopelagic and a few are even concentrated at upper bathypelagic depths. Catches off the Canary Islands have given intriguing indications of such vertical stratification, especially of how the three most abundant species of *Lensia* seem to have partitioned the water column.

Siphonophores, especially the smaller species of diphyids, are among the most abundant of the larger forms of zooplankton. They are so transparent and elusive that it takes some little time before one realizes that a catch is full of the smaller kinds.

In their natural surrounding they must be invisible. When they luminesce, a bluish glow lights up the glassy bells of the colony. The purple sail or by-the-wind sailor and its relatives which, like the Portuguese man-of-war, are members of the plueston, were once classified as siphonophores but are now given their own order.

The units of the colony hang from a circular or oval, gasfilled float that is strengthened by chitin and in *Velella* bears and upright sail. There is a large central digestive member surrounded by many reproductive members that produce free medusae and in turn are fringed by a cycle of prey-catching members. Like *Physalia*, the chondrophores are purple to deep blue in colour and have their sailing headquarters over warmer oceanic regions.

Nearly all the deep-sea species of the jellyfish class seem to lack a fixed polypoid stage in their life history. Nearly all belong to the order Coronatae and range from flattened to domelike forms that vary in diameter from a few to 25 centimeters.

The ordinal name refers to the groove that crowns the umbrella. Just below this coronal groove the jelly of the bell is fashioned into a circlet of thick pedalia, some or all of which bear a single solid tentacle. The margin of the bell is scalloped into lappets that alternate with the pedalia. As Buchsbaum and Milne say. "The beautiful sculpturing of these masses of jelly remind one of some of the gelatin desserts that have been shaped in grooved, dome-like metal moulds."

The predatory powers of *Perlphyllz hyacintha,* which has a dome-like, purplewashed bell and 12 tentacles, has already been stressed. Hardy remarks that this species is usuallyillustrated `... with its tentacles hanging limply down: in life they are extended in all directions feeling outwards for possible prey'.

The saucer-like umbrella of *Atolla* has brown to dark red markings and bears two alternating sets of pedalia with *16-32* tantacles. The disc-like *Nausithoe rubra* is a 'beautiful dark madderbrown'. Representatives of these and other coronate genera are commonest at mesopelagic levels in the subtropical belts but range into colder regions, where they may be taken at epipelagic levels.

In the other groups of jellyfishes, *Pelagia noctiluca* with purple-rose or lilac colouring, is one of the few species that are truly oceanic. It is mainly in epipelagic form, though like the deep-sea species, luminescent. Boxjellies and many-mouthed jellyfishes are predominantly inhabitants of warm inshore waters but certain species live in the upper waters of the open ocean.

Kramp makes special mention of the 'beautiful brown and violet *Crambionella orisini,* a rhizostome jellyfish that was often taken by the Danish Research Vessel *Galatbea* during her passage from Africa to India. Females of the pelagic octopus, *Argonauta,* the paper nautilus, sometimes attach themselves to this jellyfish.

In Kramp's words, Between Mombasa and the Seychelles we saw the remarkable sight of a number of *Argonauta* ensconced on top of jellyfish, *Crambtonella orsini,* which drifted in large numbers round the ship. To the best of our knowledge this had never been seen before.

Ctenophora (Comb-Jellyfishes)

Comb-jellyfishes are organized in much the same was as cnidarian jellyfishes. The outer and inner tissues are separated and supprted by a massive gelatinous matrix that is strengthened by connective tissue fibres and shapes the organism.

As in jellyfishes, the meogloea contains amoeboid cells, but there are also muscles fibres, and presumably nerve cells. The two groups are also alike in that the only body space is the endodermlined digestive cavity, which has a mouth but no anus.

This similarly in degree or organization and the disposition of the body plan in radial symmetry is recognized by placing the two groups in the grade Radiata. The other grade of animals with definite tissues and organ systems, which comprises most of such animals, are bilaterally symmetrical and called Bilateria.

More precisely, the ctenophores are biradially symmetrical Radiata. Most species have mobile tentacles that are densely studded with adhesive cells; cell unique to ctenophores and quite unlike the nematocysts of cnidarians. One ctenophore, *Euchlora rubra,* has nematocysts but no colloblasts. The protruding heads of the adhesive cells are very sticky and cling to prey.

At their inner ends they are attached to spirally coiled contractile filaments that yield to the pull of struggling prey but cannot be wrenched loose'. Ctenophores without tentacles just swallow their prey. There are about 80 species, all marine. No ctenophore is built for propulsion by jet reaction—most species make their slow way by the sequential beating of ciliary plates that are disposed in radial rows over the animal.

Each ciliary plate is made of numerous long cilia that are fused at the attached end and so look like the teeth of a comb, hence Ctenophora, meaning '*comb-cearers*'.

The play of light on a moving ctenophore may be the only indication of its presence. As Buchsbaum and Milne say, 'The delicate transparency of comb jellies makes them all but invisible in the water, so that often they reveal themselves only in the rippling iridescence of the rows of beating combs as they diffract the light.'

At night, though, ctenophores are brilliantly luminescent, and they flash along the comb rows when distributed by the sea. Lighthill implies that movement by ciliary activity is appropriate only in small organisms. At first sight, ctenophores seem to be conspicuous exceptions, for species that move by beating ciliary comb range in size from a few millimetres to about 20 centimetres.

First, though, the beating is by sizeable ciliary paddles not by tracts of discrete cilia. Moreover, the paddles beat in effective metachronal rhythm, more precisely in antiplectic metachronism, which differs from symplectic metachronism in the lag in beat between one ciliary unit and the next is greater, and indeed sufficient to give each paddle greater freedom of movement during its effective stroke.

Studies of the ciliated protozoan *Paraueclum* and the ctenophore called the sea gooseberry showed that this kind of metachronal rhythm is more effective than the other in propulsive power. Even so, most ctenophores are slow swimmers. But members of the tentaculate group Lobata, such as *Ocyptosis maculata,* move quickly by strong down beats of their large and wing-like oral lobes. Venus' girdle a ctenophore that is compressed in the tentacular plane to a beautiful band-like form and reaches a length of 1.5 metres, swims by undulations of its body.

Apart from a few bottom-dwelling forms, ctenophores are pelagic animals par excellence. Species such as *Beroe cucumis* are cosmopolitan in the widest sense and in cold waters may be tinged with pink or lavender colours. The oceanic species are predominantly epipelagic but there are two deep-water genera.

Tunicata

Much as pelagic cnidarians have evolved, as it were, around their hydrostatic skeleton to become predominant among gelatinous forms of carnivorous zooplankton, so have pelagic tunicates attained comparable success among gelatinous forms of herbivorous zooplankton.

There are over 2,000 species of tunicates, most of which are ascidians and benthic in habit from intertidal to deep-sea levels. Since the organization of pelagic cnidarians is easily run, it is understandable that some species live at deep sea levels well below the productive surface zone. Though they may be taken below this zone, and there are a few deep-sea species, pelagic tunicates are primarily forms of neritic waters and the epipelagic zone of the open ocean.

Again this might be expected if one considers their dependence on species of phytoplankton. Like their benthic relatives, pelagic tunicates trap such food by means of mucous secretions that come, as already stated, from the endostyle, which is a glandular part of the pharynx.

Larvacea are so called because in certain respects they are like the larval stages of ascidians. Most of the eighty odd species are tiny and transparent with a trunk barely exceeding 1 millimetre, and a tail usually not more than three times the length of the trunk. Special cells on the forward part of the trunk rapidly secrete a transparent and gelatinous filter-chamber (house) that varies in size and complexity from species to species.

The house is much larger than its maker, which draws one water through the filters by lashing its tail. In most species of family the water enters through a fine sieve and then circulates inside through a very fine filtering system that funnels to the animal's mouth.

Inside the mouth the food particles are trapped in a conical sieve of mucus that is maintained and formed by endostylar and ciliary activity. A twisted cord of mucus and food enters the stomach. During feeding water pressure builds up in the house and eventually a hinged door by the exit to the house suddenly opens to release a jet of water, and as one jet follows another these extraordinary animals move among the finest pastures of the sea.

Species of another family secrete a food trap that is pushed by the animal when it is feeding suspended outside its filter. By such means larvaceans trap the smallest kinds of phytoplankton down to one micron in size.

Larvaceans are very common in all parts of the ocean. Species diversity is greatest in subtropical waters. More than is realized at present, these minute tunicates must be an important source of food for young stages of deep-sea animals and for vertically migrating mesopelagic animals, such as lantern fishes.

The largest larvacean. *Bathychoradeus,* which reaches a length of more than 77 mm and builds a very voluminous house without filtering windows, is a deep-sea (*mesopelagic*) form. *Bathychordaeus,* which is believed to feed on detritus and small planktonic organisms, has an enlarged sac-like *oesophagus* and stomach.

Pyrosomids, salps and doliolids form the other group. (Thalicea) of apelagic tunicates. Inside their gelatinous tunic, they resemble ascidians in their general organization, but are modified for pelagic feeding and locomotion in that the atrial cavity is behind the pharynx and discharges water through a posterior opening.

Pyrosomids develop into colonies'... consisting of numerous individuals embedded in the wall of a gelatinous tube, which is usually long, more or less cylindrical, closed at one end and open at the other. The zooids have their branchial siphons at the outer surface of the tube, the atrial siphons opening into the common cloacal cavity.

Each accordingly passes a water current from the exterior to the interior, where the individual currents fuse into a powerful stream emerging from the open or posterior end of the tube. Locomotion is accordingly a continuous jet propulsion. A diaphragm at the open end narrows the aperture and increases the velocity.

Colonies are often colourless and transparent, but may have a slightly yellowish, blue-green colour. They may be soft and gelatinous or rigid and cartilaginous in texture. The outer surface is never quite smooth'.

Though predominatly warm-water forms, pyrosomids are founn in all oceanic areas apart from the Arctic Sea. Colonies range in length from a few centimetercs to over 4 metres. As their name suggests, pyrosomids are luminescent: indeed, they may well produce the brightest light of all luminous marine organisms. The light comes from photogenic bodies on each side of the phorynx and has been said to range in colour through red, orange, greenish, sky blue, greenish-blue and azure.

But one set of careful measurements showed that most of the light energy is blue-green with a secondary greenish peak. Luminescence has also been observed in the larvacean genus *Oikopleura* and in salps and doliolids.

Though they are rather simply organized, doliolids have a complex life cycle. Berrill (1950) remarks that their '...method of attainment of a sexually mature form is probably the most intricate in the whole of the animal kingdom'.

This sequence of events is less complicated in salps. Essentially both groups have an asexual and a sexual phase in the life history. In salps solitary individuals with no gonads (oozooids) have a budding stolon producing a chain of individuals that form the aggregate phase and produce ova and sperm.

The fertilized eggs produce oozooids. Doliolids have solitary sexual individuals (gonozooids) and the fertilized eggs produce tailed larvae that grow into asexual oozooids. Each oozooid has a ventral budding stolon and a dorsal spur at the posterior end.

Buds formed by the stolon migrate up and become

attached to the dorsal spur where they grow into multiple lateral rows of feeding individuals (trophozooids) and amedian row of phorozooids that eventually become detached and grow into gonozooids.

In both groups and sexual forms have an individual structure, but in essence the means of propulsion and food procurement are similar. Propulsion is centred in the muscle bands that entirely or partly encircle the body wall just below its substantial gelatinous tunic.

As these muscles relax and contract water is taken into the brancbial chamber and expelled in jets from the atrial chamber. Bursts of locomotor activity are interspersed with longer and irregular pauses.

Much as in larvaceans, phytoplankton organisms and particles that enter the pharynx are trapped in a mucous net that is secreted by the endostyle and drawn across the pharynx by ciliary activity. In the aggregate form of the salp *Pegea confoederata,* for instance, the mucous net is conical and continuously renewed and is evidently able to retain particles as small as 1 micron.

The remains of flagellates, coccolithophorids, diatoms, tintinnids, foraminiferans and radiolarians were found in faecal material. Salps are larger than doliolids. Solitary salps vary in length from a few to 200 millimeteres.

Chains of aggregate forms may reach lengths of 2 metres or more. Single doliolids span from a few to 30 mm and the chains of zooids on the nurse oozooids reach 40 centimetres. Like other pelagic tunicates, salps and doliolids are mainly cosmopolitan in the subtropical and tropical belts.

A number of species extend into temperate and polar waters. In equatorial regions much of the planktonic biomass in the summer months consists of salps, doliods and pyrosomids.

Sometimes the sea is thick with tunicates and at right the light displays of pyrosomids is a sight not to be forgotten. Apart from turtles, fishes and numerous invertebrates, notably heteropods, their main predators are various kinds of seabirds.

Pelagic Snails (Pteropods and Heteropods)

In two groups of pelagic gastropods the muscular foot, which is one of the key features that contributes to the outstanding success of gastropods in benthic habitats, has been modified into wing-like appendages. Hence they are known as pteropods.

More precisely, the wings are outgrowths of the sides of the foot and form the propulsive organs of the shelled pteropods (Thecosomata) and the naked pteropods, which both belong to the opisthobranch group of gastropod molluscs but are not closely related.

In one family of shelled pteropods (Limacinidae) the wings and head emerge from a transparant shell measuring a few millimetres in length or width. Morton (1958) writes, 'The foot has a broad sole and bears an operculum, and from its sides are produced long parapodia, or "wings", muscular at the bases and thin towards the tips.

Lmmacina rows itself upwards in a broadly spiral course, using these as oars, and drops as a dead weight by holding them together motionless above the body.' Species of the family Cavoliniidae have a bilaterally symmetrical shell, which may be long and conical *(Cresets),* dorso-ventrally flattened *(Clio)* or inflated and flasklike *(Cavoli$_r$ia).*

Cavolinta tridentata, which is about 22 mm long, bears a pair of greenish, leaf-like appendages, each strengthened by a central vein that emerges from a rear angle of the shell.

Presumably these appendages enable the animal to plane through the water, and not lose too much height, between bursts' of swimming activity. Members of another family are large, gelationous and close to perfect buoyancy.

They have no true shell but lie in a boat-like pseudoconch made of chitin. 'In *Gymbulla* a pair of broad based wings are attached along the sides.

They are flapped like a butterfly's wings with an up and down rowing movement; and they show also an undulation from the front backwards, which provides a forward thrust.

In *Gleba* and *Corolla* the pseudoconcha is a wide saucer and the wings have grown together in a heart-shaped fringe which is moved gracefully in swimming like a skirt or the bell of a medusa' Morton, 1958. The shelled pteropods feed on micro-organisms, particularly on the minute plants of the nanoplankton.

Species of Limacinidae and Cavoliniidae set up ciliary currents on the wings and mantle, so drawing in organisms and food particles that are trapped by a film of muscus. Cymbuliids secrete a large mucus web. *Gleba cordara,* which has a wing span of about 50 mm, spins a 2 metre-span web by means of mucus glands at the tips of the wings.

'Gently manipulated by the cilia in the animals' elongated mouth, the free-floating membrane slowly sinks, collecting a variety of tiny planketers on its sticky surface. Eventually, *Gleba* draws most of the web down its proboscis and consumes it'.

Like other members of the zooplankton that arc predominantly herbivorous, the shelled pteropods are both common and abundant. In the Beaufort Sea region of the arctic, the cosmopolitan species *Spiratella belicina* appears to be the cause of a sonic scattering layer at a depth of 50 metres.

Nearly all pteropods are epipelagic and most species live in the warmer oceanic belts, where a number are circumtropical. Shells of pteropods may be found in the sediments, but pteropod ooze occurs in significant amounts only in the Atlantic Ocean, mainly in the central South Atlantic.

Cymbuliids, with their large digestible cast-nets of mucus, are evidently very efficient herbivores and probably much more abundant than is appreciated at present.

The naked pteropods are much less abundant than the shelled species, which is not surprising in view of their carnivorous habits The body is more or less spindle-shaped and may attain a length of more than 40 mm. Gymnosomes move quickly on powerful wings.

'The animals move upwards or forwards by sculling with them, twisting them at the "wrist" to allow a power stroke at both the upward and downward beat'. The prey, much of which consists of shelled pteropods, is taken by a formidable buccal armature consisting of various combinations of prehensile hooks, sucker-studded tentacles and adhesive oral papillae. One species, *Clione limacina,* which, like its prey *Spiratella helicina, is* sometimes an important item in the diet of baleen whales in arctic waters, is also similar in its 'bipolar' distribution.

But gymnosomes are most diverse in warmer oceanic regions. Heteropoda, known by some 25 species, belong to the large prosobranch group of snails. Like many other planktonic animals they are virtually transparent.

In one family comprising the smallest species, the coiled transparent shell is so compressed and keeled that the snail stays upright as it swims, which, as in all heteropods, is effected by undulations down a fin-like part of the foot. The shell is much reduced in the Carinariidae and entirely absent in the Pterotracheidae.

Members of both families have an elongated gelatinuous body that is close to neutral buoyancy. When these heteropods swim quickly, which is aided by sepentine movements of the body in pterotracheids, the fin side is uppermost.

All heteropods have large tubular eyes moved by special muscles, which may well help them find their prey. Morton (1958) writes, "The fast swimming Heteropoda, with movable telescopic eyes, are active predators, catching medusae, small fiish, copepods, and pteropods and themselves falling prey to larger heteropods.

The large Pierotrachea and Carinaria carry the buccal bulb at the end of their flexible "trunk" periodically thrusting out the sharp-toothed radula to seize prey, or even to attack the human finger-tip. Heteropods swallow their prey entire, which also includes hyperiid amphipods, salps and arrow-worms.

The headquarters of these lively pelagic snails is in the epipelagic zone of the subtropical and tropical belts, where numerous species are more or less cosmopolitan.

But their place in the ocean is limited by comparison with that of the arrow-worms, which are also predators and have much the same range of lengths.

Chaetognatha (Arrow Worms)

Three groups of worm-like invertebrates are predominant in marine eco-systems. In the zooplankton of both the neritic and oceanic provinces, the Chaetognatha or arrow-worms are much more ubiquitous and abundant than the pelagic species of polychaete worms, which, together with the nematode worms, are outstanding among the worm-like invertebrates in benthic environments.

Arrow-worms maintain their place in the ocean, which goes back to Cambrian times, with relatively few (about 50) species. Most are below 40 mm in length and the entire range is from a few to over 100 millimetres.

A rounded head, armed on each side with a set of grasping spines, passes evenly into a transparent, torpedo-shaped body that extends in the horizontal plane into one or two pairs of laterial fins Wand a terminal tail fin. The fins are immobile and supported by a double set of delicate fin rays.

Apart from a few benthic species of the genus Spadella, all arrow-worms are planktonic and most species are placed in the genus Sagitta. In most arrow-worms the body is more or less stiff and turgid and is held straight between periods of activity. Movement comes from the quick contractions of substantial bands of muscles that run down the body.

Arrow-worms seem to spend much of their time slowly gliding on their fins... eventually however, they begin to sink, whereupon they resort to swimming to maintain their level. Swimming consists of short swift forward darts, each

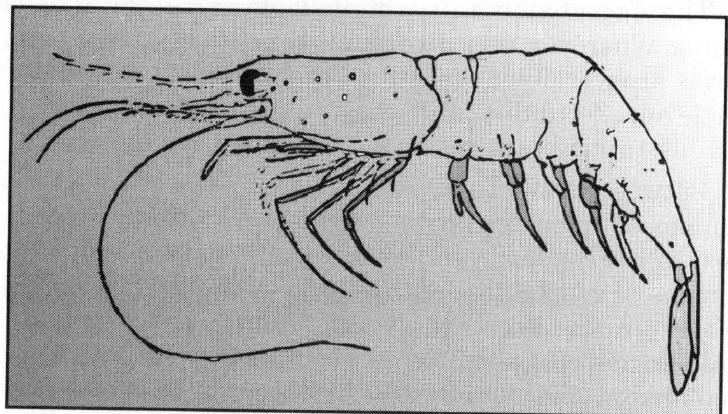

Figure 2.10: A deep-sea prawn, Systellapsis debilis (5-6 cm).

covering a distance of around 5 cm; this is followed by gliding on the momentum of the darts and a return to floating.

The formard dart is so rapid as to be difficult of analysis but is believed to result from the alternate contractions of the dorsal and ventral longitudinal muscle bands of trunk and tail. This would seem to produce up-anddown body waves and tail flirts, but the movement is so rapid that it appears to the naked eye like a trembling'.

Arrow-worms may dart to elude their predators or take their prey, both of which are detected and located by groups of bristle-bearing sensory cells that remind one of the neuromast organs in the lateral line system of fishes Marshall, 1972. The prey is seized in the spine-like jaws and swallowed whole.

Copepods are the most frequent prey, but other chaetognaths, euphausiids, medusae, pelagic tunicates and young fish are also taken. Perhaps chaetognaths feeds on most kinds of zooplankton that they are able to swallow.

Unlik most other groups of planktonic animals, chetognaths have no larval phase. They are hermaphr-odites and the ovaries mature after the coelomic cavities of the tail are full of ripe sperm. In Sagitta self-fertilization is said to be common.

The eggs batch into young looking rather like the adults. Wherever they live, and in oceanic waters arrow-worms are world-wide in occurrence at epipelagic and deep-sea levels, fecundity and the cycles of reproduction are such as to maintain relatively abundant populations.

Between New York and Bermuda the displacement volumes of epipelagic arrow-worms were usually second to those of the copepods and averaged 13 per cent to the total volumes of zooplakton. At levels from the surface to 1,000 metres in the western North Pacific the biomass of arrowworms was about 10 per cent of the total biomass of zooplankton and micronekton irrespective of the depth.

In absolute terms the number of individuals-and the number of speciesper unit volume of water are greater in the upper 200 metres than below this layer. Species diversity is greatest in warmer oceanic waters. Thus, in the subtropical and tropical waters of the Indo West Pacific faunal province, which extends from the Red Sea and east coast of Africa across the Indian Ocean to the Hawaiian Islands and Polynesia in the western Pacific, about 85 per cent of the known species have been recorded, some of which are circumtropical.

Species such as *Sagitta bipunctata, S. enflata, S. bexaptera* and *Pterosagitta araco* are cosmopolitan at epipelagic levels in both temperate and warm oceanic waters. Species with populations centred at mesopelagic levels include *Eukrobni'a spp. Sagitta lyra, S. decipiens* and *S. macrocephala. Bathyspadella edentata,* which may be batbypelagic, is without jaw spines.

Certain mesopelagic species may be tinged in varying degrees with orange or red pigments. Except for the bluish pigmentation of neustonic forms, other chaetognaths are virtually transparent.

Diurnal changes of level associated with vertical migrations are well marked in certain species in neritic waters, where populations tend to be concentrated near the surface during the night. Oceanic species may be partial migrators or nonmigrators. In the Southern Ocean, where

Sagitta gazellae is one of three endemic species. there was no clear evidence that *gazellae* is a diurnal migrator.

But there are seasonal changes of level associated with the life cycle. David's analysis suggests that maturing animals gradually sink and eventually release their eggs at about a depth of 1,000 metres. The eggs rise and probably batch at about 250 metres, the young arrowing into juvenile animals that mass in the upper 100-metre layer.

These growing populations migrate to the 100-250-metre layer in the winter months, but rise and resume life in the surface layer in the spring.

POLYCHAETE AND NEMERTEAN WORMS

Like the arrow-worms. planktonic specics of polychaete worms are predatory and known from relatively few (about 60) species. Most belong to four families (Alciopidae, Tomopteridae, Typhioscolecidae and Phyllodocidae) of which the first two are entirely comprised of pelagic forms and best known to marine biologists. Fauvel (1959) has caught their attraction.

'Les Tomopieris et les Alciopiens pelagiques sont transparents commedu crystal et serainent a peu pres invisibles n'etaient less taches vivement colorees des parapodes des premiers *et les enormes yeux rouges des seconds.'* ('The pelagic tomopterids and alciopids are as transparent as crystal and would be almost invisible if it were not for, firstly, the vividly coloured spots on their parapodia and, secondly, their huge red eyes.')

Some of the brightly coloured spots on the parapodia of tomopterids are traceable to rosettes of yellow cells that luminesce through large surrounding hyaline cells. The large projecting parapodia end in paddle-like expansions, and it is by sequential strokes down the parapodial series that tomopterids row swiftly through the sea.

Alciopids have shorter (uniramous) parapodia but have lively motions as they seek their zooplankton prey, aided no doubt by their large camera-type eyes. Most species of pelagic polychaetes live in warmer oceanic regions and as Friedrich (1969) remarks, ... they often occur sporadically and solitarily, so that the role of these predatory forms in the biology of the pelagial is of secondary importance in comparison with that of the benthonic polychaetes which play a very important part in the biology of the benthos'.

One species, *Tomopteris septentrionalis,* lives at epipelagic levels in artic and antarctic regions but at deeper levels in the warmer oceanic belts. Certain deep-sea species of *Tomopteris,* as Sir Alister Hardy (1956) illustrates so well, are almost entirely red.

Some of the pelagic nemertean worms, all of which seem to live at deep-sea levels, also develop red pigmentation. The body, which ranges in length from 6 to 20 cm and is usually widened, sometimes with undulating edgcs, has a gelatinous mesenchyme.

Species of *Pelagonemertes* have a transparent body and a branched, orange or red-coloured gut others, such as *Bathynemertes hardy* I, are scarlet or plumcoloured. *Nectonemertes* has been observed to swim rapidly with strong undulations. Pelagic nemerteans presumably use their long protrusible proboscis to take their prey in the zooplankton.

Nekton (and Micronekton)

Nektonic animals have greater swimming powers than those that comprise the zooplankton, and much as one may usefully distinguish a micro-zooplankton from the larger forms of zooplankton so may the smaller kinds of nekton be considered as micronekton.

The members of each group have a size and activity range that may be sensed by the dimensions, mesh size, and towing speed of the nets that have been designed to catch them most effectively. A small conical net with a mesh size of 200 microns and towed at a speed of about half a metre per second is used to sample the micro-zooplankton.

A larger conical net with a mouth area of I square

metre (70 cm diameter) and mosh size from 300 to 500 microns, which is usually towed at something like 1-3 metres per second, samples the larger members of the zooplankton.

The net designed by the Institute of Marine Resources. La Jolla, California, to sample the micronekton of eastern Pacific waters has a mouth area 2.3 square metres and oblong meshes of 5.5 × 2.5 inm.

Towed at about 5 metres per second, this net takes the larger crustaceans, small cephalopods and fishes of a size range from about 1 to 10 centimetres (Strickland, 1972).

Nets such as the Isaacs-Kidd tidwater trawl and the modified Tucker net used by the Institute of Oceanographic Sciences with a mouth area of several square metres, and smallest (codend) meshes of 77 meshes-inch, are towed at speeds of 1i knots (0.77) metres per second) or more and catch micronekton and smaller fish and squid members of the nekton (> 10 cm in length).

Still larger mid water nets increase the size range and catches of nekton, but there are still limitations. As Clarke (1971) observes, 'Large commercial midwater rawls are regularly used for biological sampling. and here the mouth is held open by otter boards floats on the headline and large weights on the footrope.

Examples are the British Columbia midwater trawl, the Engels midwater trawl and the four-door Cobb midwater trawl. These trawls are difficult to operate from multi-purpose oceanographic vessels.

However, they are valuable in sampling animals up to about half a metre in length. The catch is not easy to use in comparative studies because the mouth area and shape change with speed, and graduation of the mesh size along the net makes interpretation difficult.

Even with nets of this enormity with mouth areas of 2,800 square metres and 750 square metres (less when fishing because they close up due to drag) the larger animals, known to be present by observation are not caught.' Such observation may be made by examining the food eaten by sperm whales.

Elsewhere, Clarke has calculated that a sperm whale of eleven tons eats no less than 130 tons of squid in a year. Many of these squid will be over a metre in length and some, such as *'Architeuihis,* span well over 10 metres.

Apart from their size and activity, micronektonic animals differ from those of the zooplankton in form and organization. The shrimp-like myside and euphausiids and the deep-sea prawns, like the squid and fishes, have more or less streamlined spindle-shaped bodies. In keeping with their size and complex organ system, al these animals have elaborate gills.

In the twilight world of mesopelagic levels eyes are very well developed and many species have complicated constellations of light organs. Size, form at d activity are correlated with massive muscular systems. The joint exoskeleton of crustaceans is associated with a complex system of muscles, which are very highly developed in micronektonic forms.

In a mysid, euphausiid or deep-sea prawn the weight of muscle must be well over 50 per cent of the total weight. During cruising movements the most active part of this mus-le is that moving the swimming legs (pleopod), but during escape movement the whole of the powerful abdominal musculature propels the animal backwards by a series of rapid tail-flips.

Muscle and skeleton make up much of the dry weight of such forms. For instance, the dry weight as a percentage of the wet weight reaches 15 per cent or more in amphipods, mysids and euphausiids. In siphonophores, medusae, arrow-worms and tunicates the corresponding figure ranges form 0.3 to 10.0.

The mantle muscle of squids, which expresses propellant jets from the mantle cavity; is about 30 per cent of the body weight in active species. In active fishes the segmented lateral muscles used in swimming are 40 per cent or more of their total weight. Lastly, a high percentage of micrronektonic forms are carnivores, whereas many zooplankton species, forming much of the total biomass, are herbivores.

In othee words, compared to the zooplankton most organisms of the micronekton may be regarded as predatory-behaviour-machines that are quite expensive to maintain. The greatest contrast between the micronekton and nekton is in the epipelagic zone.

Her is the swim of baleen whales, sperm whales, dolphins, turtles whale-shark, basking shark, isurid in gray sharks thresher-sharks marline, spearfish sailfish, swordfish, tuna fling-fishing and various squid (ommastrephid and onychoteuthid).

This is Isaac's vivid appre ciation, 'The pelagic region contains some of the largest and most superbly designed creatures ever to inhabit this earth the exquisitely constructed pelagic tunas; the multicoloured dolphinfishes, capturers of flying fishes; the conver ationals porpoises; the shallow- and deep feeding swordfishes and toothed whales, and the greatest carnivores of all, the bateen whales and some plankton-eating sharks, whose prey are entire schools of krill or small fishes.

Seals and sea lions feed far into the pelagic realm. In concert with these great predators large carnivorous sharks await injured prey. Marine birds, some adapted to almost continuous pelagic life, consume surprising quantities of ocean food, diving, plunging, skimming and gulping in pursuit.

Creatures of this region have developed such faculties as advanced sonar, unexplained senses of orientation and homing, and extreme olfactory sensitivity.' Compared to all these large and active members of the nekton, the crustaceans and small fishes of the micronekton are indeed shrimps and so much whitebait.

One need only contrast this state of affairs with

conditions immediately below in the mesopelagic zone which is the most productive part of midwater space, to see how much life is scaled down in the deep sea.

If the size range of micronektonic animals is taken to be about I-lo cm, then virtually all of the larger crustaceans and many of the cephalopods and fishes are included. Some of these animols, such as certain species of euphausiids, prawns, squid and lantern-fish. swim up to the epipelagic zone each day, where they form part of the micronekton during the night.

The nekton of the mesopelagic zone consists of,a few large species of fishes and squid. Though more is known now of the larger forms of pelagic deep-sea animals, Hjort's description of this fauna as Lilliputian still seems apt.

Euphausiacea

If their near ubiquity, trophic range and numerical abundance are considered together, the euphausiid shrimps emerge as the 'copepods' of the micronekton. Considering just the oceanic species, euphausiids live from epipelagic to bathypelagic levels throughout the four oceans.

Species fall into three main trophic types, as in copepods herbivores (whe conditions *are* favourable), omnivores and carnivores. In one measure of abundance, the mean displacement volume of euphausiids was about 5 per cent of the total catch figure in a series of net hauls in the epipelagic zone between Vew York and Bermuda.

As already stated the corresponding figure for copepods was about 50 per cent, and one must always remember that copepods are much easier to catch than eughausiids. At levels between the surface and 1,000 metres from stations in the western North Pacific, the biomass of euphausiids was 8 to 15 per cent of the total catches of zooplankton and micronekton.

During the antarctic summer, vast swarms of euphausiids, nourish baleen whales, crab-eater seals, penguins and fishes. The 85 species of euphausiids (order Euphausiacea) are closely related to the large order of decapod crustaceans which includes shrimps, prawns, crayfish, lobsters and crabs. Indeed, certain authorities class them with the Decapoda.

All species are shrimp-like in appearance and hane well-developed antennules and antennae that bear olfactory and tactile receptors. Just behind the mouth the paired mandibles, maxillules and maxillae collect, macerate and manipulate the food just before it is swallowed.

The outer branches of the 6-8 pairs of thoracic limbs, to which the gills are attached. produce currents of water from which food particles are screened by the setae on the inner branches (endopodites) of these limbs. Euphausiids cruise by the sequentia paddling motions of the two pairs of swimming limbs on each of the first five segments of the abdomen.

The sixth and last segment has no such appendages but articulated at the end of this segment is a slender tapering telson, on either side of which is a flattened

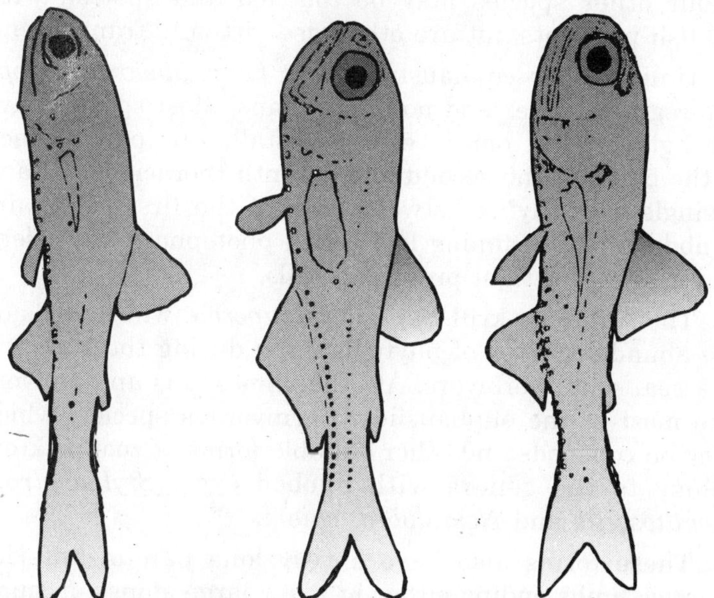

Figure 2.11: Stomiatoid fishes. From top to bottom.

biramous limb (uropod). Telson and uropods form a tail fan, which is brought powerfully into play by a series of quick flexures of the muscular abdomen, when the animal shoots backwards, perhaps to evade a predator.

Species range in length (tip of rostrum to end of telson) from about 10 to 150 mm. The largest of the epipelagic species is *Euphausla superba* the whalers' krill of oceanic areas south of the Antarctic Convergence.

Smaller species of *Euphausia* and various smallish species of other genera (e.g. *Stylocheiron* and *Nematoscelis)* are typical of the warmer epipelagic waters Other species of these three genera and representatives of such genera as , *hysanopoda* and *Nematobrachion* live at mesope!agic levels. Large species of *Thysanopoda* (adults of *T. cornuta* and *T. seinicaudata* may exceed 100 mm in length) and *Bentheuphausia amblyops,* which are widely distributed, are typical of bathypelagic levels.

These species of *Thysanopoda* may be largely red in colour other species may be touched and spotted with reddish pigments but are otherwise virtually transparent.

Unlike other euphausiids, *Bent heuphausia amblyops* has regressed eyes and no light organs. Most species have ten light organs: one in each eye stalk, one pair at each of the bases of the second and seventh thoracic limbs and a single organ lying between each of the first four pairs of abdominal swimming legs. Each photophore has a lens and a group of tight-producing cells.

The Antarctic krill *Euphausia superba,* which feeds on the abundant crops of phytoplankton during the summer, is a sea. sonal herbivore. At other times it is an omnivore like most of the euphausiids. Carnivorous species, which prey on copepods and other suitable forms of zooplankton, belong to the genera with bilobed eyes: *Stylocheiron, Nematoscelis* and *Nematobra. chion.*

These forms also have a very long pair of anterior thoracic limbs ending either in a few large spines or small chelae, presumably for grasping captured prey, The functional design of the feeding and and alimentary devices of euphausiids and the part these animals play in oceanic food chains and economy are considered later.

Each day certain mesopelagic species migrate between their daytime levels and the surface water where they find much of their food while other species, both epipelagic and mesopelagic, stay at one level. Recent survey shows that migrant species of the same genus overlap in their daytime vertical ranges. whereas there is vertical segregation between nonmigrant species,

Euphausiids produce-small eggs and after fertilization each eventually hatches into an ovoid nauplius larva a millimetre or less in length. The nauplius has three pairs of swimming legs which later form the two pairs of antennae and the mandibles.

The nauplii develop after moulting their exoskeleton into second stage nauplii. A series of growth stages and moults takes the nauplius through a metanauplius stage calyptopis stages (with a hooded carapace and more definite abdomen) and furcilia stages, which become more and more like the adult.

The last furcilia stage moults, forms and changes into an adolescent. Deep sea species apparently prawns at depth and the sequence of events in some mesopelagic species, at least, is that the early larval life is passed in the productive surface water. If all the larval stages of bathypelagic species remain within the depth range of the adults one wonders how the smallest larvae manage to feed themselves.

Deep-sea Prawns

Deep-sea prawns are often prominent in the catches of large nets fished from epipelagic to bathypelagic levels in temperate to tropical reaches of the ocean.

In biomass the catches of such prawns are close to those of the euphausiids and ganerally comprise about 10 per cent of the total biomass in samples taken between the surface and a depth of 1,000 metres.

Deep-sea prawns, euphausiids and fishes make up most of the micronekton and their contributions respectively range from 15—25. 35-50 and 20-45 per cent of the total biomass. As a rule euphausiids are smaller than prawns, so the former easily outnumber the latter.

Moreover, the part played by prawns in the pelagic life of subpolar and polar waters is insignificant compared to that assumed by the euphausiids. Shrimps and prawns form the Natantia, or swimming decapod crustaceans.

Lobsters, cray fish and crabs, which are predominantly benthic. The characteristic features of natant decapods are the body is laterally compressed; the base of each antenna is articulated to a movable, plate-like scale, the thoracic legs are slender. and there is a full complement of well-developed swimming legs (pleopods) on the abdomen.

In nearly all species the carapace juts forwards as a prominent rostrum, which is often saw-toothed and compressed. During quick movements, when prey may be seized, the compression of the body and rostrum will tend to keep a prawn from veering off a line that is guided by the antennal scales, which act as rudders.

Compensatory movements of the scales may also help to control rolling and pitching. Deep-sea prawns thus seem to be well fitted for an active, pelagic life. They also have a much lighter, less well calcified exoskeleton than the bottomdwelling, crawling decapods.

The latter never have a laterally compressed body, antennal scales are vestigial or absent, and the pleopods which are not used in swimming, are often reduced.

Swimming decapods fall into two main groups, Penaeidea and Caridea, which are readily distinguished. Carideans have expanded sideplates (pleura) on the second abdominal segment that overlap those of the adjacent segments, whereas there is no such overlap is in penaeids.

Furthermore, most penaeids have a small claw at the end of the third thoracic leg, which is not developed in carideans.

Deep-sea penaeids such as *Gennadas, Funchalia* and sergestids have extremely long whip-like antennae and the lash is thickly and regularly set with setae, many of which are presumably tactile or olfactory in function.

These prawns are rather slender in form but they have powerful pleopods and in some the rearmost pairs of thoracic legs have become strong jointed oars. One has the impression of sudden powers to dart on prey detected by the searching sweeps of the long sensory antennae.

Deep-sea carideans also have strong swimming legs. Species of the family Oplophoridae are stoutly made: those of *Notostomus* have an 'inflated' carapace.

There are over 200 species of pelagic prawns, most of which range from 10 to 100 mm in length. The largest reach lengths from 130 to 180 mm and often seem to live. near the deep sea floor (Omori, 1974). Nearly all species live in the subtropical and tropical belts and some extend into temperate waters.

As already indicated, deep-sea prawns are poorly represented in the coldest oceanic regions, where, for instance, in antarctic waters five species have been recorded. TI ere are parallel differences in the latitudinal diversity of euphausiids, but the six antarctic species, especially *Euphausia superba,* are key forms in the economy of life.

Euphausiids are also more prominent than the prawns, both in individuals and species, in the temperate to tropical reaches of the epipelagic zone. Of the mesopelagic prawns, those that live around 500-700 metres or above by day are transparent, or translucent with scattered red chromatophores, or half red *(Sergestes spp., Funchalia* and *Parapandalus).*

All species that occur below 500-700 metres by day (lower mesopelagic) have numerous small chromatophores and a heavily pigmented exoskeleton that ranges from deep red to scariet in colour. Many of the mssopelagic species undertake daily vertical migration but in bathypelagic forms (e.g. *Acanthephya stylorostralls, Hymenodora glacialis* and *Physetoearis mlcrophthalma)* the main part of the population lives below 1,000 metres throughout the entire day.

Certain bathypelagic species seemingly extend down to depths of 5,000 metres or more. Some deep-sea prawns have light organs. The most remarkable are modified liver tubules (Organs of Pesta) which may have reflectors and lenses to enhance and concentr ate the down-welling light. These organs are developed in both penaeids and carideans but in the former group, for instance, they occur only in the semi-transparent upper mesopelagic species.

The lower mesopelagic species of *Sergia* are a uniform red colour and there are no Organs of Pesta. Dermal photophores may also be present, and those that have lenses also live at upper mesopelagic levels.

The all-red species develop dermal photophores without lenses and these live, as already stated, at lower mesopelagic levels, *Bathypelagic species*, suco as *Sergia japonicus,* are without photophores.

There are thus correlations between type of pigmentation, photophore development and depth ranges. Indeed, as Foxton has suggested, and is considered elsewhere these correlations remind one of parallel developments in certain cephalopods and fishes.

Deep-sea penaeids produce numerous small eggs that are shed into the water. After fertilization they hatch into minute nauplius larvae with the usual three pairs of swimming legs that later become the antennules, antennae and mandibles. The nauplii change after moults into protozoea stages, which have functional mouthparts.

Then comes a zoea stage (called acanthosoma in some sergestids) in which some or all of the thoracic limbs function in swimming, but the abdominal limbs are absent or rudimentary. The succeeding post-larval stages look more and more like the adults and the abdominal segments acquire their full complement of setose swimming legs.

In some species of *Sergestes* and *Sergia* the carapace of the protozoea and zoea larvae bristle with long spines that may have a protective function as well as helping to slow down the sinking rate of these larvae.

Whereas the eggs of sergestids are a fraction of a millimetre in diameter, deep-sea carideans produce relatively few, Jarge eggs which range from about I to 4 mm (along the longest axis). The eggs are carried by the female on long setae of the abdominal limbs where thay stay until they hatch into relatively advanced stages. After a few post-larval stages the adult form is attained.

The early larval stages are passed within the egg, and as Omori says, the advanced young that are produced have well developed swimming ar d feeding powers that give them considerable safety from predators and independence of supplies of minute food.

The penaeids rely on safety in numbers. A comparative survey of such reproductive and larval strategies in the deep-sea will be found elsewhere.

Peracarid Crustaceans

In euphausiids, shrimps, prawns. lobsters, crayfish and crabs, which are classed as eucarids, the carapace is large and covers the thorax. The eggs, if carried, are attached to the pleopods. Peracarids, as implied by their name, have a brood pouch in which the female bears the rather large eggs until they hatch.

The carapace, if present, is not fully developed and leaves at least four of the thoracic segments free. Pelagic peracarids belong mainly to the opossum shrimp group and the Amphipoda. Cumaceans, tanaids and isopods, which are entirely or largely benthic.

Deep-sea mysids have been taken in nets fishing below *6,000* metres. But they are commonest at mesopelagic levels, where live species of such genera as *Gnathophausia, Eucopia and Lophogaster*.

Species of *Gnathophausia,* which are red and look like prawns, are also prawn-like in size, Most mysids are from 20 to 30 mm in length, but a female *Gnathophausia ingens* measured 351 mm (from tip of rostrum to tip of telson) and other large specimens of this species range from 155 to 210 mm.

Species of *Eucopia,* which have long, frail-looking thoracic legs, are also red. There is a fine painting of a scarlet *E. unguiculata* by Sir Alister Hardy Many mysids are filter feeders, but the deep-sea forms are likely to be omnivores. Amphipods have sessile eyes, no carapace, and in nearly all species the body is flattened from side to side and presents an arched profile. Representatives of the subgroup *Hyperiidae* live at epipelagic and mesopelagic levels over most of the ocea. \ *Hyperia, Ihemisto* and *Pararhemisto* species are most common are pipelagic levels, and individuals of one live commensally it seems, with large jelly fishes.

The transparent frail-looking Diogenes shrimps *(Phronima)* with enormous eyes on a hammer head, have more drastic designs on gelatinous organisms.

Adult *Phronima.* which may be common at mesopelagic levels and range from about 20 to 30 mm in length, will take over the swimming bell of a siphonophore or the test of a salp or pyrosomid and then proceed to eat away the inside of such structures until a gelatinous barrel is left.

The *Phronima* lives in the barrel and moves it along by limb movements that draw water through its shelter, which also serves as a brood chamber and a nursery for the young. Cystisoma, another fragile but larger (15 cm) amphipod with a near crystalline transparency, is seen from time to time in nets fished at upper mesopelagic levels. Of one fine photograph, the Angels remark, 'The only colour in this specimen is the interference colour of the eyes and some red pigment from the animals' last meal.'

But certain mesopelagic hyperiids are all red, and again we are indebted to Sir Alister Hardy for fine paintings. Representatives of the largest group of amphipods (Gammaridea) also form part of the mesopelagic fauna. Even so, the peracarid crustaceans are less prominent in the economy of pelagic life than their eucarid relatives, the euphauriids and prawns.

Cephalopoda

When a research vessel is on night station, squids and fishes may be attracted to the illumination cast into the sea by the deck lights. The squid may also be drawn to the fishes, though it is difficult to tell predator from prey. One sees dark spindleshaped bodies and glinting eyes moving just below the surface.

Even in brighter surroundings a darting squid may be momentarily mistaken for a fish. Through such illusions, an observer ismade a ware of the extraordinary similarities between cephalopods and fishes. Beside the streamlined shape of active species and their prominent camera-like eyes, the two groups resemble each other in numerous other respects of their gross and fine structure.

These resemblances form just one striking illustration of the brute fact that during their evolutionary history distantly related kinds of plants and animals have converged in functional design in response to the pressures of natural selection in a given medium, or to subdivisions of a medium ranging from broad adaptive zones to narrow ecological niches.

Though this is not the place to consider convergence, it must be said that this aspect of evolution is prominently displayed in the deep sea. Many of the 700-800 species of modern cephalopods live in the deep sea. Pelagic species are mostly squids together with a number of octopods, a few sepioids (cuttle-fish) group and the vampire squid. Most of the octopods and seploids are bottom dwellers.

Epipelagic species, notably the ommastrephid squid, have firm torpedo-shaped bodies and strong swimming powers, which reside in their thick, muscular mantles. One species and a hooked squid, which have large triangular fins, have power enough to dart out of the water and fly through the air.

After take-off, flying-squid are sustained largely, it seems, by continued expulsion of jets from the funnel, and they may cover distances of 50 metres or more. When a small shark approached two *Ommastrephes,* both were observed to turn a dark red and shoot out of the sea to a height of 2 metres.

Though primarily epipelagic, ommastrephid squid have been photographed by baited cameras at mesopelagic levels. As in fishes epipelagic species are considerably larger than most of their relatives of lower levels.

The flying-squid (*O. bartrami)* reaches a length of about 1 metre, while *Dosidicus gigas,* "an ommastrephid of Chilean waters, grows to well over 2 metres.

There are smaller and more delicate squid in the epipelagic zone, especially members of the family Cranchiidae. Species of *Cranchia andLiocranchia* which may be quite abundant, have soft, globular bodies. large eyes, short arms and rather long tentacles. Apart from certain internal organs, these squid are virtually transparent.

The liver is a particularly opaque organ, but in *Liocranchia* it is silvered over, presumably to hide it from predators. Cranchia scabra attains a body length of about 50 mm, which is less than half the length of an adult *Locranchia reinhardti*.

Concerning the cephalopods of deeper levels, closing mid-water nets fished accurately from *RRS Discovery,* the research vessel of the Institute of Oceanographic Sciences, UK, have enabled Malcolm Clarke and his colleague C. C. Lu to determine the vertical ranges of the commonest North Atlantic species throughout their growth stages and where possible to trace the daily changes in level of the species that undertake vertical migrations.

They also found a latitudinal gradient in species diversity. From 60° N to about 20° N the number of species taken roughly doubled for every ten degrees of latitude. Thus, like other groups of marine organisms, cephalopods are most diverse in the warmer parts of the ocean. Much of what follows is taken from their work.

Mesopelagic squids of the upper 700 metres '.. are in

general small squids rarely exceeding 30 cm in overall length excluding 30 cm in overall length excluding the tentacles and averaging only about 15 cm with large numbers adult at 4-5 cm'. In the North Atlantic the commonest small squid of mesopelagic levels include *Abraliopsis affinis: Histioteushis spp.; Pyroteuthis margaritifera* and *Mastigoteuthis schmidti. In Abraliopsis* and *Histioteuthis* most of the numerous light organs are on the lower side of the animal.

The few jewel-like light organs of *Pyroteuthis* are set around the eyes, at the tips of its tentacles and in the body. Indeed, an outstanding feature of mesopelagic squid '.. is the possession of numerous small to large light organs or photophores distributed chiefly upon the vertical surface of the mantle, funnel, head, arms and the ventral periphery of the eyeball'.

Many mesopelagic fishes also have numerous light organs along their underparts, and, as in squid, some of these organs are of elaborate design. Again, mesopelagic squid and fishes have large and very complex eyes, but no fish has eyes such as develop in histioteuthid squid, which have the left eye much larger than the other. Two small seploid squid, *Spiruia* and *Heteroteuthis*. are also common in parts of the Atlantic at mesopelagic depths.

Spirula floats head downwards, buoyed by an internal, gas-filled spirally chambered shell, and shining upward at the end of the body just above the shell is a light organ. *Spirula is* the only deep-sea cephalopod with a gasfilled buoyancy system.

In diverse squid neutral buoyancy is achieved by buoyancy chambers filled with relatively light ammoniacal liquids. Such a system is found also in certain deep-sea prawns. Fishes manage well, if need be, with a gas-filled swimbladder. The other sepioid, *Heteroteuthis,* looks rather like its bottom-dwelling relatives.

The ink it discharges on being disturbed is mixed with a luminous secretion, which reminds one of the clouds of luminous material emitted by certain deep-sea prawns and fishes. As already implied, convergent evolution is rife in the deep sea, especially in visual, luminescent and buoyancy systems.

In the catches from their North Atlantic stations, the two commonest midwater octopods studied by Clarke and Lu were *Vitreledonella richardi* and *Japetella diaphana,* which both grow to a mantle length of about 60 mm. Both species are virtually transparent and have the gelatinous consistency and something of the appearance of a jellyfish.

The first is mesopelagic at all stages of growth, whereas half-grown and adult *Japetella* were taken at upper bathypelagic levels. A third gelatinous midwater octopod, *Amphitretuspelagicus,* has upturned tubular eyes that are reminiscent of similar organs in certain mesopelagic fishes.

The finned octopods, such as *Cirrothauma* and *Girroteuthis* once thought to be bathypelagic, have been photographed near the deep-sea floor. They are also gelatinous and the large webs between their arms make them look even more like jelly fishes.

The paddle-like fins project from a relatively small boy. *Cirrothauma is* blind and certain species of *Cirroteuthis* have regressed eyes. No deep-sea octopod, or indeed any kind of octopod, is know to be luminescent.

The vampire squid, *Vampyroteuthis infernalis,* which is more or less cosmopolitan between lower mesopelagic and bathypelagic levels, is neither a squid nor an octopus. With its eight broad-webbed arms, paddle-like fins and soft gelatinous bodyi t looks rather like a finned octopod, but there are certain outstanding differences.

Actually the vampire has ten arms, two being modified into long, mobile sensory filaments that are withdrawable into special pockets. Thus, it is also quite distinct from the squids in which two of the ten arms develop into special, prey-catching tentacles. *Vamp) roteuthis is* placed in a separate order called the Vampyromorpha.

It is also unlike any octopod in its well-developed set of

light organs. Besides the many small organs that stud the skin, there are two clusters of luminous nodules on the back of the 'neck' and two large lights, each with an 'eye-lid.' just behind the fins. Males reach a length of about 125 mm. females about 200 mm.

Some kinds of mesopelagic squid migrate daily between their daytime levels and overlying waters, where they swim by night, and presumably find much of their food. Among the species that are most common the North Atlantic, Clarke and Lu found reasonable evidence that *Abrallopcis affinis, Bathyteuthis abyssicola* and *Pyroteuthis margaritifera* appear at epipelagic levels during the night. Other common squid, such as *Heliococranchia pfefferl* and *Mastigoteuthis schmidti,* are nonmigrators, as are the octopods *Japetella* and *Vitreledonella.*

There are also brtogenetic migrations during the life history. The catches from *Discovery* stations suggest that most deep-sea cephalopods start their young life in the epipelagic zone and assume deeper levels as they grow to maturity. There is uncertainty as to where the eggs are laid.

In *1895,* when the *Princesse Alice 1,* the research yacht of Prince Albert 1 of Monaco, was working near the Azores the local whalers harpooned a sperm whale, which happened to die under the bottom of the yacht. In its death flurry it spewed up parts of great squids new to science.

On returning to Monaco the Prince had his yacht fitted as a whale-catcher and began a series of investigations on the sperm whale and its food, showing that it feeds mainly on squids, some with tentacles as thick as a man's arm and studded with great suckers bearing catlike claws.

But for these and subsequent investigations, and for the occasional stranded specimen, our knowledge of squids, particularly of their size range, would be very restricted. Sperm whales feed on medium-sized to the largest squid, which must be more abundant and larger than appears from the cathes of mid water nets.

FISHES

Small fishes, euphausid shrimps and deep-sea prawns make up most of the micronekton. Why are small squids not a major constituent, or are sampling devices at fault? Perhaps squids are the hardest to catch. through the elusive powers of a fish would seem to be much the same as those of a squid of the same size and shape. But the smallest pelagic squid are larger than the smallest fishes.

Most of the prolific species of midwater fishes, which nearly all belong to the stomiatoid and lantern-fish groups, range from 25 to 70 mm in length when mature. If the arms are included in the length, hardly a small sized species of squids falls into the above size span, which suggests that the smallest squids are more elusive than the smallest fishes.

But are the latter more productive than the former, bearing in mind that animals feeding nearest the primary producers in a food chain are the most productive.

Are small fishes better able than their squid counterparts to take advantage of the wealth of small (mainly copepod) food in the ocean ? Small stomiatoids and lentern-fishes certainly feed on eggs, larval stages, copepods, ostracods, arrow-worms, euphausids and so forth, but less is known of squid food.

Prey caught by a squid or octopus is conveyed by the arms to the beak-like jaw and there chewed, if necessary, into pieces small enough to enter the oesophagus, which is restricted by its passage through the brain and cephalic cartilages. Diverse predatory fishes are able to swallow prey up to half of their own size.

Small species of midwater squid, including migratory species, also feed on copepods and euphausids and certain mesopelagic and bathypelagic cephalopods have suckers small enough to grip small planktonic animals. Further observation and investigation is needed.

In the meantime one suspects that the fishes are able to take a wider range of small prey organisms, and thus feed closer than the cephalopods to the photosynthetic base of the food pyramid. Sor instance, it is improbable that any genus of deep-sea cephalopod has species to match the abundance and ubiquity of those in the stomiatoid genus *Cyclothone.*

The most diverse mesopelagic fishes, both in numbers of species and individuals, are stomiatoids and lantern-fishes. There are over 300 species in thn second group and about 250 in the first. Both groups have a few representatives at bathypelagic levels near 1,000 metres and below, where, indeed most fishes belong to the dark-coloured species of *Cycloihone.*

But in diversity alone the bulk of bathypelagic fauna consists of ceratioid angler-fihes, which are known from about 100 species. *Stomiatoid fishes,* which seem most closely related to the smelts, typically develop four rows of prominent and complex light organs along the underparts of the head and trunk and two rows at the same level along the tail.

Small silvery kinds, such as *Vinciguerria. Valeneie-nnellus* and *Bonupartia Gonosstomatidae*) and some of the *hatchet-fishes* (*Sternopty-chidae*) are common at upper mesopelagic levels, but they are usually outnumbered by small transparent fishes, again of the genus *Cyclothone.*

The smaller species of stomiatoids have relatively long jaws bearing numerous pointed teeth, part of a prey-taking system to deal with the range of food organisms already listed. Most of the larger species belong to the families Astronesthidae, Chauliodontidae, Stomiatidae, Melanost-omiatidae.

Idiacanthidae and Malacosteidae, and they all live at various levels in the mesopelagic zone. These fishes also have large jaws and some of the foremost teeth are fanglike. Tine-like teeth on mobile pharyngeal bones at the back of the throat grip and convey sizeable prey to the gullet.

Such prey may consit of euphausiids, prawns, squids and fishes, and the gullet, stomach and wall of the abdomen are distensible enough to accommodate large meals. If the predator is stealthy and quick enough food may be a single fish or squid that is at least half the length of its captor.

Besides the ventral rows of light organs, the larger and rapacious kinds of stomaitoids have large photophores below and behind the eyes. The latter cheek lighst may be very large and in some species better developed in males.

Many species have thousands of small lights all over the skin and in the clear gelatious tissue that may invest the body. Part from *Chauliodus spp.,* nearly all species have a chin barbal bearing beads and inlay of luminous tissue that may be very elaborate. In *Chauliodus* the second ray of the dorsal fin is very long. has a luminous tip.

Perhaps these are angling devices with a luminous bait. The species of *Chauliodus* and *Siomlas,* which are serpentine fishes with a protrusible upper jaw, are silvery-sided with a honey comb scale pattern, framing, at it were, the body lights. The largest species reach lengths of 300 mm or more.

The black or bronze-coloured astronesthids, which range from about 40 to over 300 mm in length, have a conventional fish form and a fin pattern much like that of a smelt. The melanostomiatids are much more diverse in form, ranging from long slim types through fusiform types to short squat types.

Female idiacanthids copy, as it were, the black skin and serpentine from that is so common in melanostomiatids, but the males are not only pale-coloured but much the smaller sex. Though also black, the malacosteids have evolved in other directions. The suspension of the jaws is so angled backwards that the gape is very wide.

There is no floor to the mouth cavity and the bare lower jaws with their strong pointed teeth remind one of the lethal mechanism of a rat-trap. The rat-trap fishes are also unusual in that the cheek photophores produce red light, which is true also of *Pachystomias*, a *melanostomiatid.* All other luminescent organisms emit blue-green light.

Like the smaller kinds of stomiatoids, lantern-fishes are

so successful that a sizeable net towed at mesopelagic levels almost everywhere in the ocean is almost bound to catch them. If, as seems very likely they have not colonized the Arctic Sea, then it is improbable that other groups of midwater fishes will have gained a footing.

All lantern-fishes are more or less spindle-shaped and bear much the same fin pattern. The jaws bear rows of small teeth, and though some kinds are larger and wider mouthed than others, there is no marked contrast between small planktoneaters and larger predators on the micronekton, such as obtains in stomiatoids.

While large lantern-fish are naturally able to accommodate somewhat larger food organisms than small species, the diet of all kinds is much as that already stated. The entire size'range in the family is from about 25 to 250 mm.

While the number and disposition of the lights on the underside of the head and abdomen is more or less standard throughout the group, the constellation of lights that may be seen on a fish in profile seems to be unique in each of the 250 odd species. When lantern-fishes display these lights, are the individuals of a particular species exchanging recognition signs. There are also differences between the sexes in the development and placing of certain ,luminous organs.

Most of the organs on the head and body look like minute pearlbuttons. The structure and function of lantern-fish lights are considered elsewhere.

If the stomiatoids have evolved from smelt-like ancestors, their nearest deep-sea relatives are the argentinoid fishes, which are related to the salmonoid groups. The argentinoids are small mouthed fishes of diverse body form ranging from short squat types through conventional types to elongated types.

Apart from *Argentina* and its near relatives, which live near the bottom over slope areas, these are fishes of the mesopelagic zone. Their adaptation to this twilight world is evident in their large and sensitive eyes, which are considered elsewhere.

Here we need only note that tubular eyes are prevalent in this group. Such eyes, with parallel optical axes that may , be directed upward or forward, have also evolved in certain stomiatoids, two families of alepisauroid fishes, giganturoids and the free-living males of the ceratoid angler fish genus, *Linophryne.*

In *Opisthoproctus* and certain other deep-sea argentinoids there is a well-developed anal light organ containing luminous bacteria. A unique king of light organ has evolved in fishes of the family Searsidae, which are also mesopelagic and have large and very sensitive eyes.

Each organ of a pair, which is set on the shoulder region above the pectoral fin, forms photogenic cells in a darkly pigmented sac that opens backwards through a pore. In *Searsia,* at least, these shoulder organs discharge a bright cloud of luminous particles.

Some species have other luminous organs, most of which are set on the underside of the fish. Apart from *Platyetrotes* and *Platytroctagen,* which have a compressed and keeled body, searsids are fishes of fusiform appearance.

The closely related aleopcephalids are much more diverse in form and fin pattern. For instance. *Aulostomatomorpha* has a long tubular snout, *Leptoderma* an elongated body with a long anal fin. *Xenodermichthys a* fairly elongated compressed body and *Alepocephalus is* more or less fusiform with opposed and set-back dorsal and anal fins.

Species of *Alepocephalus,* which may reach more than 0.5 m in length, live near the bottom at depths down to 2,000 metres or more. Some alepocephalids are mesopelagic to bathypelagic. Genera such as Xenodermichthys have light organs while others have none.

There are over 1,000 species of deep-sea midwater fihes and at least one half are stomiatoids and lantern-fishes. The next most diverse groups are the deep-sea anglers and alepisauroids.

Lanternfishes and alepisauroids are placed in the same order, but they are very different in functional design and habit. Most of the 100 odd species of ale pisauroids belong to the family *Paralepididae*, which have been called barracudiaas because some species remind one of those predatory fishes par excellence, the barracudas.

Indeed, alepisauroids, which live from mesopelagic to upper batbypelagic lelels, seem to prey largely on other fishes; and they are well equipped to make a living. Whether spindle-shaped or long-bodied, and there is a marked trend in the group for the evolution of elongated types, the jaws and inner parts of the mouth bear large stabbing teeth or a set of prominent pointed teeth paralepidids.

The dentition consists also of many small, pointed teeth that help to grip the prey. Like the predacious stomatoid fishes, alepisauroids have an expandable sac-like stomach, and even in species with body walls of limited distensibility prey up to one-third of the length of the swallower can be accommodated.

Nearly all species are non-luminous. Certain of the alepisauroids are among the largest species of midwater fishes. Alepisaurus ferox, the lancet-fish, grows to about 2 metres Anotopterus to about 1 metre and the largest barracudians to about 50 centimetres. The main range is from a few to 30 centimetres.

Though larger than most lantern-fishes, few alesauroids undertake daily vertical migrations that so occupy so many of their relatives. Indeed, the alepisauroids and other sizeable midwater fishes that are adapted to take large meals seem to be non-migrators.

Most of the migrators are fishes of modest size with moderate capacities for relatively small food organisms. If, as seems very likely, vertical migrationsare fooding sorties, are the migrators 'compelled' upwards to find the best feeding grounds or are there other factors? Arc the non-migrators able to stay at one level because of their trophic oppertunism? Giganttiroid fishes and the gulper-eel Saccopharynx are also adapted to take a wide size range of food organisms. The giganturoids are elongated to long slim fishes with a wide mouth, a fearsome dentition and large forwardly-directed tubular eyes.

Their scientific name refers to the very prolonged lower lobe of the tail fin. Like many other mesopelagic fishes, they are silvery. The species of Saccopharynx and the related genus Europharynx are very elongated, dark-coloured fishes of lower mesopelagic to bathypelagic levels. Much of their length is taken up by a long rapering tail that is tipped with luminous tissue.

The well named *Eurypharynx pelecanoides*, with extremely long jaws that remind one of a pelican's bill, grows to a length of about three-quarters of a metre: the species of Sacca pharynx reach lengths of between 1 and 2 metres. Though Eurypharynx has a much larger mouth than *Saccopharynx*, the *latterhas* a *moreormidable dentition*, a much larger stomatch and a longer abdominal cavity.

Indeed, *Saccopharynx* can be swallow very large fishes but Eurypharynx feeds on relatively small prey, such as the young stages of prawns and small fishes. Though the *gulper-eels* are highly modified fishes the fact that *Eurypharynx*, at least, produces transparent, leaf-like *leptocephalus larvae* may well mean that the true eels are their nearest relatives. Midwater eels include the *snipe-eels*, the bobtailed snipe-eel and *Serrivomer*.

The snipe-eels have long, pliable beak-like jaws, set with many minute teeth, and a very elongated body that may reach a length of nearly a metre. They have relatively large, wide-open eyes, where as the bobtailed snipe-eel, like other bathypelagic fishes, has small, reduced eyes. *Cyema*, which has a short tail '*arrow-feathered*' with opposed dorsal and fins, may reach a length of about 125 mm.

The whale-fishes, which also have small or reduced eyes, are also bathypelagic. Whale-fishes have large, small-toothed jaws and a considerable capacity, expandable if need be, for copepods and other small zooplankton. One *species*, *Barbourisa rufa*, which grows to a span of at least 20 centimetres, is one of the few raddish-coloured midwater fishes.

The other darkcoloured whale fishes are sometimes tinged with red. Like other midwater species, whale-fishes are widely dis. tributed in the ocean, but they are not commonly caught. Out of some 10,000 fishes representing well over 100 species that were taken off the *Canary Islands*, *Badcock* found two *specimens* and *species* of *whale-fishes*.

When the whale-fishes have been properly revised they may well prove to be as diverse as the Melamphaidae, a family with which they have some affinity. *Melamphaids*, with their round snout, short body, large mid-set dorsal fin and bands of minute teeth in the jaws remind one rather of *lantern-fihes*.

They also have about the same size range. But their *thoracic pelvic* fins, each with a spin and 6 to 8 jointed rays, are part of a character complex that relates them to the 'lower' spiny-finned fishes. Melamphaids, which are mesopelagic to bathypelagic fishes with brownish to dark colouring, live in all major regions of the ocean, except for the Arctic and Mediterranean Seas.

They are without light organs. In one genus there are 'dwarf' and 'grant' species. The former are evidently adapted for life in the impoverished central water masses, the letter for more productive peripheral waters.

Fish faunas of neritic and coral-producing seas are dominated by 'higher' spins-finned fishes, but representatives of this great teleost group are in a minority in the deep sea. Midwater forms include the giant swailowers and the trichiuroid fishes, predatory types with marked piscivorous ways. *Chiasmodon niger* can swallow fishes larger than itself.

The trichiuroids are diverse in form, ranging from elongated types, such as *Paradiplospinus, Gempylus,* the snake mackerel and *Aphanopus,* the black scabbard fish, to those with a fusiform body, e.g. *Ruvettus,* the oil fish and *Nealotus.* Trichiaroids, which are somewhat reminiscent of alepisauroid fishes in form, fearsome dentition and sleek appearance, are represented also by a number of large species.

The snake mackerel grows to about 1 metre, the black scabbard fish to 1.25 metres and the oil fish to over 1.5 metres. There is a deep line fishery for the second species off Madeira. The depth of transition from the twilight of the mesopelagic zone to the sunless waters of the vast bathypelagic zone is near 1,000 metres.

Dark-coloured species of *Cyclothone,* whale-fishes, gulper-eels and the deep-sea angler-fishes, which related to the angler-fishes or monkfishes and frog fishes, form most of the bathypelagic fish fauna. Angler-fishes bear a rod and bait on top of the head, but in the ceratoids it is developed only in the females.

In all but one form the bait is luminous and differs in structure from species to species. In each species the individuality of the lure may be seen in its shape, photogenic complex and Iressing with dermal appendages, but there is no evidence yet that particular lures attract particular kinds of food organisms.

It seems more likely that female angler-fishes, which have a wide, well armed mouth and distensible accommodation, are trophic opportunists, able to take a wide size range of prey. Perhaps the males are prawn to a proper partner after receiving the 'right' sign stimuli from her lure. But the males of a few species have tiny eyes, though their olfactory organs, like those of most male ceratioids are hig ly developed. Perhaps the females with keen-nosed mates lay scent trails during the breeding season.

The males are much the smaller sex and their dwarfing is but one aspect of a complex and trenchant sexual dimorphism. After metamorphosis the males are about 10-40 mm in length, and there is little subsequent growth.

They are active more or less fusiform fishes but in nearly all species the females are tubular or pear-shaped.

In certain ceratoids the male becomes parasitic on the females and in such forms she is from three to more than ten times the length of her mate. The free living males of *Himantolopdus,* which is one of the ceratoids that seem to have other mating arrangements grow to about one-thirteenth of the length of the largest known female.

Though their means of reproduction are unsual, the lifehistory patterns of deep-sea angler-fishes are much like those of most other mid-water fishes, whether mesopelagic or *bathypelagic*. Their eggs, which must be very numerous but have yet to be found, are probably shed in the depths and develop as they float up Nards.

During metamorphosis to adult form there is descent to the levels of the adult living space. Here a few of the young fishes will eventually reach maturity,

Zooplankton

This comprises representatives of almost every major group of animals, either as adults or as larvae. Of the 135000 species of bottom-dwelling invertebrates, 1,00,000 have a freeswimming larval phase which lasts on average about two to four weeks. Just as there were single-celled forms of phytoplankton, so there are single-celled members of the zooplankton.

These are not capable of producing their own food but are consumers relying upon ingestion of other organisms, usually smaller phytoplankton and bacteria. Some Protozoa produce elaborate shells or tests and include the radiolarians and foraminiferans.

Globigerina is an example of the latter and enourmous numbers occur in warmer seas and when they die, accumulate on the ocean floor. This forms *Globigerina* ooze, a sticky glutinous sediment covering large tracts of ocean bed e.g. in the Tasman Sea *Coelenterates.*

Both holoplanktonic and meroplanktonic species exist. In the class Hydrozoa there are two main phases to the life history-bottom dwelling colonies of hydroids which give rise to meroplanktonic reproductive bells such as *Bougainvillea ramosa.*

The common jellyfish *Aurelia aurita* which has a world-wide distribution, does not have the usual sessile hydroid stage. Larvae produced from its reproductive bell, settle and then bud off new larvae known as ephyrae, rather like a pile of saucers.

Ctenophores

These include the common sea gooseberry ; they are holoplanktonic and unlike true coelenterates have no stinging, cells. They swim by beating eight rows of 'comb-plates' Larval herrings form an important part of the diet of the sea gooseberry *Plaurobrachia.*

Chaetognaths

Otherwise known as arrow worms these are holoplanktonic and are transparent, torpedo-shaped, active predators. Various species of *Sagitra* are found in British waters and were used by Russell in 1939 as biological indicators of different water masses.

Certain species such as *Eukrabnia hamata* tend to be found at shallow depths in high latitudes and greater depths in tropical regions to compensate for the rise in temperature of low latitude surface waters but departures from these patterns of distribution in the water column are not uncommon.

Polychaetes

A few polychaetes are permanently planktonic as adults and include *Tomopteris belgolandrca* which may reach 35 mm in length and is found in the North Sea and off Norway. Each segments of the body has lateral extentions with two branchae known as parapodia, much enlarged and flattered in these plantonic forms.

Many other polychactes are benthic and only become planktonic during the breeding season, Sexual forms are known as epitokes which have special adaptations for swimming such as the enlargement of eyes, muscular parapodia and bundles of bristles or Satae.

The activities of the Pecific Palol worm *Eunice virldis* off Samoa are now legendary. Every October and November on two days', at the last quarter of the moon there is a mass swarming of epitokous tail-ends of Palolo worms at the surface of the sea.

Unmodified head-ends retreat into crevices among the corals on the bottom. Native fishermen scoop up the swarming worms by the basketful and provide a once a-year delicacy. Some of the syllid *polychaete epitokes* such as *Autolytus* undergo an elaborate mating process with the male encircling the female with sperm threads prior to the female shedding eggs into a ventral egg sac.

Crustacean

These are by far the most numerous species, comprising about 90 per cent by weight of all the zooplankton. Holoplanktonic copepods occur in vast numbers and the calanoids predominate in most seas.

Calanus finmarchicus comprises the food of the herring in the North Sea. Copepods have adopted various strategies to avoid being eaten. There is no shelter in the open sea, hence in the surface waters most large copepods are transparent.

In deep water copepods are often red in colour but are able to hide under cover of permanent darkness. In middle depths the copepods migrate up to the surface at night where they feed in the phytoplankton-rich waters. At dawn they migrate down to avoid predators; the larger they are the deeper they need to go in order to be safe.

Species such as the amphipod *Gammarus oceanicus* are found in the open sea in the Arctic where salinity is reduced from the effect of melting ice. Further south they are found only in coastal waters where there is some dilution from rivers.

Many other crastaceans are meroplanktonic including the cirripedes or barnacles. They begin life as a freeswimming nauplius larva with two easily identifiable horns projecting laterally at the front of the carapace. After several moults a ypris larva is produced with a bivalved shell.

This stages settles and the familiar calcareous plates of the adults are secreted around the transformed animal, now filter feeding by means of the legs extended through the opening, guarded by smaller plates. The malacostracan crustaceans include both holoplanktonic forms with a dispersal phase in their life histories.

Euphausiids are widely distributed throughout the oceans and seas of the world but perhaps the best known is 'krill' *Euphausia ouperba* of the Southern Ocean. Females spawn 2000 to 5000 eggs which then sink 800 to 1000 m before hatching into nauplius larvae with three pairs of swimming appendages.

These larvae moult and graduall acquire more swimming legs and the characters of the adult. The cise to the surface by larval stages of takes a few weeks and this 'developmental ascent may be aided byupwelling currents. The decapods include many meroplanktonic forms. Penaeid shrimps such as the American Pink shrimp *Penaeus duorarum* fished in the Western Atlantic have a long life cycle.

There are several phases beginning with the nauplius, changing to a protozoea, followed by a mysis, a post larva and finally the juvenile which has most of the specific characters of the adult. Caridean prawns which predominate in European waters pass much of their early larval development in the egg membranes hatching at the zoea phase, Several moults then occur prior to metamorphosis.

The stage 1 palaemonid zoeal larva has sessile compound eyes with apparently hexagonal facets, three pairs of biramous thoracic swimming apendages, a solid triangular telson and swims upside down, tail first.

After metamorphosis the post larva has developed stalked compound eyes with square facets, lost the outer swimming branches of the thoracic limbs, developed abdominal pleopods which are used for swimming, gained the typical tail fan with lateral uropods and swims the right way up, head first. Crabs often have short life cycles with only two phases: zoea and megalopa.

Molluscs

Some complete their development to the crawling stage inside the egg, but most hatch as a free-swelling planktonic larva, which, after this dispersive phase, usually become benthic The first larval stage in the common European limpet *Patella vulgata* is a truchophore less than 0.2 mm across.

Rows of cilia forming prototroch beat and rotate through the water. Two days after fertilization the larva is transformed into a veliger with a shell and a typi cal molluscan foot. While a planktonic veliger the limpet undergoes torsion. The first 90° twist is brought about by muscle contraction.

After 30 hours full 180° torsion is complete. After 3.5 to days pelagic life ends, the velum disappears and the larval shell is lost. The edible periwinkle *Littorina littorea* ha,ches as a veliger with a 1.5 whorled shell. The trochophore stage is passed in the egg.

The pela· ˙c phase lasts for up two months and hence gives great possibilities for dispersion. Nanoplankton is wafted into the mouth by the beating of the velum cilia during the planktonic phase.

Echinoderms

Aldults are radially symmetrical but their larvae show bilateral symmetry and are usually planktonic. Eggs are shed into the sea and fertilization is external. Normally there is no brood protection but there are exceptions, particularly in polar seas.

A common larval form links the groups of echinoderms and is known as the dipleurula or 'little sides'. In the early embryo the whole surface is covered with cilia forming a closed loop. In sea cucumber larvae the band becomes sinuous. In starfish larvae the the ciliated band divides into two and again becomes sinuous. Sides of the larvae aredrawn out into long process, the three at the anterior end bearing adhesive discs. The starfish forms inside and after about two months breaks free. In sea urchins the larva is known as a pluteus and in brittlestars an ophiopluteus.

Fish

The eggs and larvae of fish are most often planktonic. But even eggs that sink or are attached to the bottom of, for example, gobies, wrasse and herring may give rise to planktonic larvae. Yolk is utilized initially until the mouth and anus become functional. Fish larvae are often quite unlike the adult and identification is difficult.

Specific patterns of pigment cells may be used before the more reliable meristic characters of ossified bony skeleton in the post larva have developed.

REPRODUCTIVE STRATEGY

In late summer and early autumn in some temperate waters such as the southern North Sea the mero plankton. Often greatly out number the holoplanktori. The reason for this dominance is that they are often produced in vast numbers.

This kind of reproductive strategy is adopted by 'r' selected species in which larg: numbers of small eggs with little or no yolk are produced. The larvae which hatch from these eggs are planktonic and often have a long pelagic life and of necessity start feeding immediately.

If there is an abundance of other small planktonic organisms then these opportunists take advantage of the food supply and many survive, some settling in new areas, the larval period allowing good dispersion. If for any reason small planktonic food organisms are in short supply, then many of the larvae die.

Within these very large broods there is considerable

potential for genetic variation. If some of the larvae meet with unusual conditions they might still be able to survive. Many invertebrates such as some polychaetes, molluscs, echinoderms and some crustaceans produce many millions of eggs in a breeding season.

Mortality is often very high. But if more than two individuals survive, the species could soon build up large propulations, dominating and possibly excluding other species from the area. This rarely happens because physical and biological controls such as competition, food supply, availability of places to hide usually prevent a species from over-running an area. These natural controls tend to maintain balanced and stable populations.

The other kind of reproductive strategy adopted by 'K' selected species involves the production of small numbers of large eggs rich in yolk. They either develop directly into miniature adults or have a very short freeswimming larval existence.

In these forms the larvae do not need to seek food immediately. There is a complete gradation between the two extremes of r' and 'K' selection. In 'K' selected species there is a greater certainty that the small numbers which are produced are well provided for and will survive.

Thus reproductive effort is either directed at large numbers of eggs poorly supplied with yolk or small numbers with more yolk and direct development. Abnormal sea temperatures may retare development. Larvae actively seek out the preferred kind of habitat, but larvae will delay metamorphosis, e.g. shrimp larvae achieve this by inserting extra moults, if conditions are not favourable.

It has been suggested that certain 'giant' caridean larvae have been carried far from their normal habitat and have continued to growwithout metamorphosing into the adult form. However, all these delays result in the larva being exposed for longer to attack from active predators or even filter feeding molluscs.

Up to 100000 larvae of a variety of marine

invertebrates may be trtrapped in 24 h at the height of the summer season by a singlemussel. Even settling may be hazardous where coeleterates lie in wait armed with batteries of stinging nematocysts.

SEASONAL VARIATION

In temperate waters increasing hours of sunlight and to a lesser extent the spring rise in temperature herald a phytoplankton bloom. This is followed closely by a peak of zooplankton grazing on the phytoplankton. This grazing and a lack of nutrients caused by a stable thermocline—a sharp boundary between warm and cold water - causes a crash in phytoplankton populations.

In autumn the thermocline breaks down under the influence of equinoctial gales stirring up the water column and bringing to the surface nutrients that had been dissolved in water too deep and therefore too dark for plant growth.

This results in a second smaller phytoplankton. bloom. In the tropics the plankton population is much more stable throughout the year. In polar seas there is no phytoplankton production during the six-month period of permanent or semi-darkness.

There is also no thermocline, so nutrients are not limiting, but phytoplankton production is confined to a brief period at the height of summer followed quickly by a single zooplankton peak.

Chapter 3

BOTTOM DWELLING ORGANISMS

Biological communities consists of groups of populations of species living together in a particular environment. Living organisms and their non-living environment are always closely iner-related and interact with one another.

The benthic or bottom-dwelling community lives on and in the floor of the seas and oceans, and ranges from kelp forests and coral reefs in shallow seas to the unusual fauna associated with soft sediments in the abyssal zone.

COLLECTING AND RECORDING GEAR

In the early years of this century Petersen designed and constructed a bottom sampler. He used it on soft sediments in shallow water and although the grab was effective, the samples were small, only 0.1 or o, 2m^2. The relative sparseness of many benthic organisms meant that the number of specimens per sample was small. Single samples therefore gave an unreliable indication of the abundance and dispersion of the fauna sampled.

The depth to which Petersen's grab sampled depended on the nature of the substrate. In soft mud penetration was good, but some of the deeper burrowing forms were not sampled at all.

Various workers have modified Petersen's basic grab in

order to overcome the problem of shallow bites on harder grounds One example is the Van Veen grab which has long scissor-like arms, clossing the grab with a lever action. Smith and Maclntyre devised a grab using the Van Veen lever system but mounted in a frame with two trigger plates. The jaws do not fire until the grab has settled on the bottom.

The jaws are then operated by strong springs that thrust the grab into the substrate, Another variation is the Holme double scoop grab which depends on a weighted frame to achieve a deep sample. These grabs all give reasonably quantitative estimates of numbers per unit area.

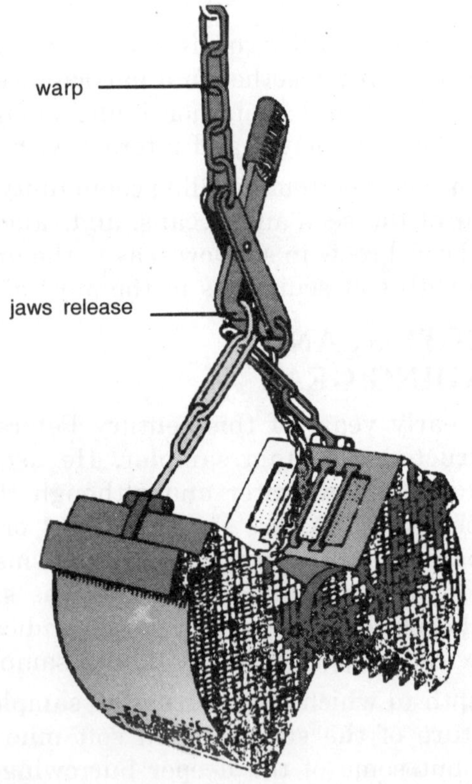

Figure 3.1: Petersen's grab. When the grab settles on the substrate, warp the goes slack and releases the jaws.

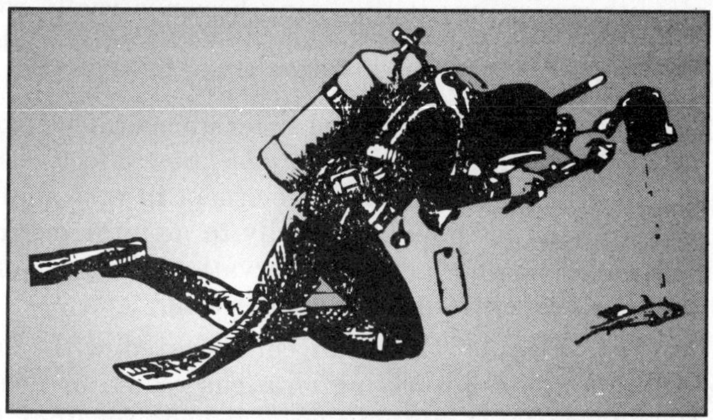

Figure 3.2: SCUBA diver using biological recording equipment including electronic flash camera and writing flate.

Grabs sample relatively small areas so there is a need to take several samples in order to obtain a representative selection of all species present. This operates on a law of diminishing returns-after an initial worthwhile increare in the numbers of species taken, subsequent grab samples produce fewer species that have not been collected already.

Another grab that is very effective at penetrating hard, compacted substrates is known as the Orange Peel grab because it consists of several 'segments'. It suffers from the disadvantage that it is impossible to estimate the area covered by the grab and the results are therefore qua-litative rather than quantitative.

The anchor dredge is a more useful means of taking a semi-quantitative sample. All of these sample the infauna or those animals living within the sediments. The other group of benthic organisms the epibenthos is collected most abundantly by dredges and trawls. Animals live on the surface of hard and soft bottom substrates.

There are many variations on the basic trawl for collecting epibenthos including a beam trawl which is strong enough to be used on rough substrates. On very rugged substrates an Agassiz trawl may be used which has the advantage over a beam trawl in that it will continue to sample either way up. If the going becomes really rough then a scallop dredge is used which looks rather like a trawl except the mesh is made of interlocking steel rings.

The collecting gears described so far all operate remotely and it is often , difficult to judge just '.how cffective they are at collecting specimens because except in very shallow, clear water it is not possible usually to monitor progress. Diving has enabled both direct observation of organisms *in situ* and closely controlled hand collecting.

In 1943, Jacques Cousteau collaborated with Emile Gagnan, an engineer work-ing on a gas regulator used in war-time Paris taxis running on coal gas. Together they designed a demand valve and the subsequent development of the Self-Contained Underwater Breathing Apparatus gave great impetus to this direct appr-oach to marine biology. Divers using aqua-lungs are able to make *unique contributions* to biological recording schemes and collecting data for this purpose has become a regular/feat-ure of many club dives.

Diving is a particularly valuable method of obtaining reliable information from hard irregular substrates where it would be difficult to operate even a steel dredge. Depths of 40 m may be achieved comfortably by competent wet-suit divers using compressed air and dives to 60 m are technically feasible, with proportionately longer stops for decompression on the way up.

On deep dives this means that the amount of useful working time on the bottom is limited. Longer dives are possible in more cumbersome hard helmet dry suits or by saturation diving. The latter method exploits the fact that once a diver has been compressed at a given depth for about 10 hours his body tissues absorb no more inert gas. The diver is able to stay at that depth indefinitely without needing extra decompression time at the end of the dive.

Teams can work in deep water for days without returning to atmospheric pressure by being transported from the bottom to the surface in a compressed diving bell and then transferring to a deck decompression chamber where they eat and sleep, still at the same pressure as the dive site on the bottom.

Spectacularly deep dives in manned submersibles such as the American *Alvin* provide a means of direct monitoring and collecting at 2500 m and more. Electronic flash photography is used increasingly by divers. This method of recording is used also in conjunction with sophisticated deep water epibenthic trawls.

Acoustic beacons relay information back to the ship to ensure that the trawl is in contact with the bottom. An angled camera, attached to the frame of the trawl is often used to take shots ahead of the net at regular intervals.

These photographic data of the benthos are necessarily restricted to the epifauna (and flora in shallower water) but they do give clear information on abundance and dispersion. This visual record is often at variance with what comes up in the trawl.

There is almost always less in the trawl than one might expect from the photo-graphs and the known distance covered as measured by an odometer. Another method of visualla recording events under orate is by closed-circuit television.

Resolution is often not as good as photographs but it does enable continuous observations of animals in their natural habitat to be made. But all these methods are expensive. However, in order to obtain a repre-sentative sample of the benthos ideally a trawl with a camera attached should be used in combination with a deep sampling quantitative grab.

Even with the array of equipment there are unexpected difficulties; as a crew becomes more adept at handl-ing gear, so the number of species caught increases. This inbuilt bias in a series of samples will be difficult to estimate.

Physical Factors

Temperature Groups of marine animals may be divided up into regions bounded by sea temperature isotherms. Animals occurring at great depth are likely to be stenothermic (able to tolerate only a narrow range of temperature) and hence would be unlikely to penetrate in to shallow water where temperature is more variable.

The 20°C isotherms to the north and south of the equator effectively delimit the distribution of coral reef formation.

Salinity Average salinity near the surface is reduced in high lattitudes from the dilution effect of ice melting. The relatively calm Sargasso Sea area at the centre of the North Atlantic gyre has a consistently high salinity of over 37 parts per thousand. Salinity becomes lowered in tropical regions only where there is a significant dilution effect from rivers such as the Amazon and Niger.

Light Visual feeders are obviously directly affected, but light also indirectly influences all benthos in that most of the food web is dependent upon p!ants fixing the energy of sunlight through photosynthesis. One of the difficulties in examining the organization of benthic communities is in finding out what powers them and keeps them going.

Below the photic zone there is very little primary production with the exception of some autotrophic bacteria. Nevertheless there are vast areas of underwater communities in complete or semi-darkness surviving without any direct energy input from the sun.

In the photic zone most of the plants are algae, either usually fixed *macroalgae* forming dense kelp forests in cool temperate waters or *unicellular* forms, mostly *planktonic* but also forming vast populations of symbionts in the tissues of coelenterates and other animals, particularly in coral areas.

There are a few flowering plants in the shallow water zone such as the eel grass *Zostera*. When Petersen was taking grab samples earlier this century he estimated that in Danish waters alone there were 24 million tonnes of eel grass growing in dense '*meadows*' in the shallow waters.

He assumed that the dead and dying vegetation from these meadows became detached and was carried by currents and tides into deeper waters and in the form of detritus undergoing microbial decomposition thus provided the source

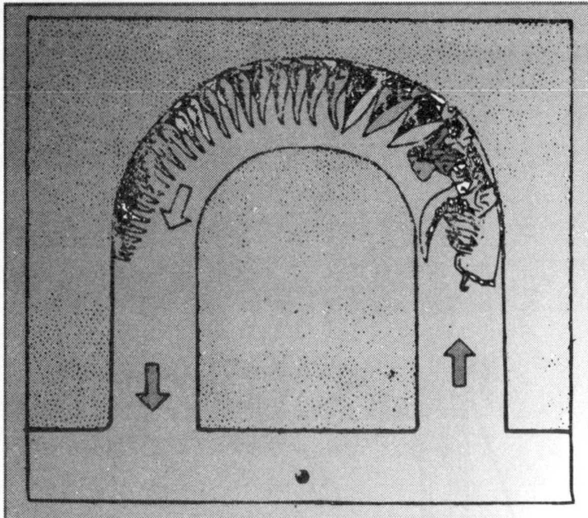

Figure 3.3: Polychaete Chaetopterus in parchment-like tube. Biological factors.

of energy for his marine benthic communities. This idea seemed sensible and went unchallenged until 1931 when there was an eel grass epidemic disease caused by a Vast areas disappeared in a few years But th fun gus benthic communities continued apparently unchanged.

It seems likely that the phytoplankton that forms the basis of the plankton food web, also forms the source of energy for the benthos as well, in the form of a continual rain of detritus from above.

Exposure and desiccation: Intertidal animals of necessity o have become adapted to continual wetting and drying out but A desiccation is a powerful barrier to the shoreward extension of sublittoral species.

Oxygen: In certain fine sediments anaerobic conditions may develop and effectively exclude many species requiring a good supply of oxygen. Many of the meiob;,nthic forms thrive in these deoxygenated conditions

Currents and tides: These will affect the dispersal of organisms especially those benthic forms that have larval stages forming part of the meroplankton.

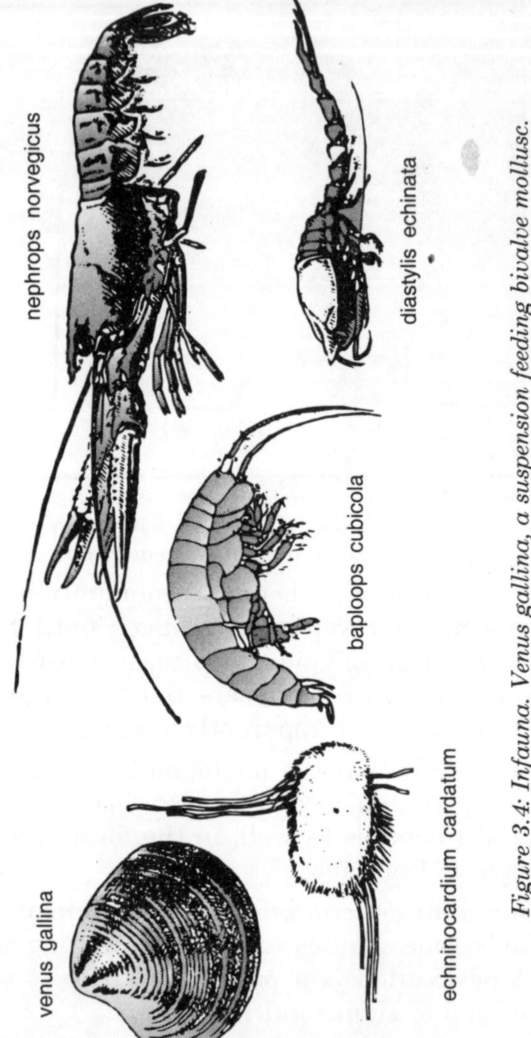

Figure 3.4: Infauna. Venus gallina, a suspension feeding bivalve mollusc.

Sediment: When a grab sample lands on deck the animals are sorted from the sediment by sieving, preserved and usually identified later in the laboratory. Before this sieving takes place a small sample is usually kept intact for sediment analysis.

After gently drying the sample it is analysed using a

stack of sieves ranging from coarse at the top to very fine at the bottom. The proportion of different grades of sediment in the sample are then worked out. Bottom sediments are often very rich in organic matter-up to 10 per cent-and this again is easily determined on an approximate basis in the laboratory.

The sample is first dried gently to drive off the moisture content, weighed, heated fiercely to burn off organic matter, cooled and reweighed. However, carbonates are inadvertently included in the weight loss of the sample. A more accurate method of estimating organic matter which excluded carbonates involved heating with chromic acid, followed by volumetric analysis.

This has now been superseded by automated carbon and nitro-gen analysers. Many physical factors exert their

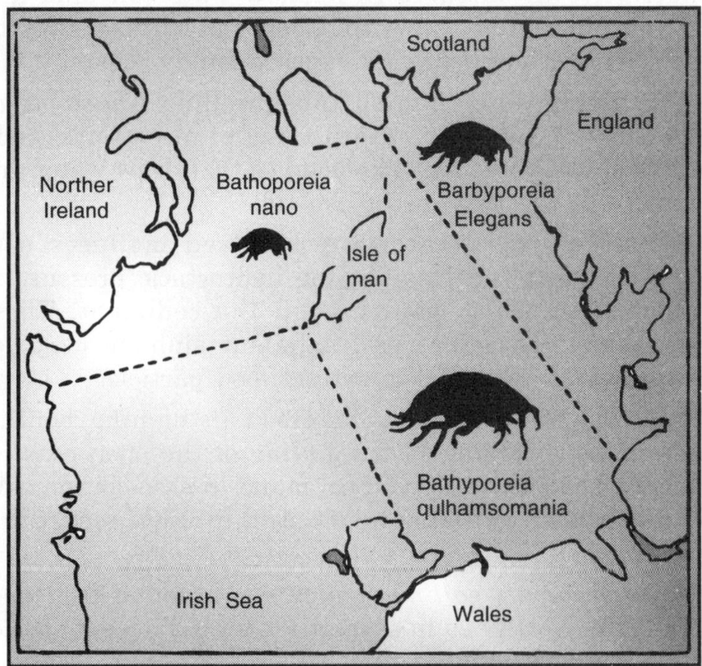

Figure 3.5: Amphipod zones in the shallow water sand of the Irish Sea.

effect through the biological properties of the organisms. Breeding seasons are often triggered by a particular temperature.

However, the relationship is usually far from direct. Other factors may delay the onset of breeding even when the threshold temperature has been exceeded. Some factors are independent of the physical environment, for example, relationships with other organisms such as a predator and its prey and also parasitism.

Competition for space can limit the number of particular species, These biological factors are much more difficult to estimate than physical factors.

Benthic Organisms-Infauna of Soft Sediments

Polychaetes: A variety of worms present in large numbers live in burrows. In shallow water on mud and sand the ubi-quitous lugworm *Arenicola marina* may be found. *Nephtys caeca* prefers exposed sandy coasts.

The highly differentiated tube-dwelling *Chaetopterus variopedatus* has the middle segments of its body specialized into three flap-like paddles that draw water and food through the parchment-like tube.

Molluscs: There are many bivalved molluscs that burrow into soft substrates using hydrostatic pressure in the foot to punch a way through the sediment. These bivalves have in-halent and exhalent siphons used for drawing water over gills to extract food particles.

Molluscs were used by Petersen in naming benthic communities including *Venus galling* of the open coast of the Kattegat and the North Sea, found in asso-ciation with *Tellina fabula* and the heart urchin *Echinocardium cordatum.*

Another *Venus gallina* community is found in deeper water in the North Sea in association with different species of molluscs and echino-derms.

Crustaceans: These form a prominent part of the benthic infauna. Adult scampi *Nephrops norvegicus* burrow in mud in depths from 5 to 70 m and eat polychaete

Figure 3.6: Life in a coral reef lagoon.

worms, small fish and other smaller crustaceans. By far the most dominant groups of benthic crustaceans are small cumaceans and amphipods. Petersen designated an amphipod dominated communit)—*Haploops tubicola* in association with bivalves.

Infaunal amphipods have been used as indicator species for communities in the shallow-water sand habitat of the Irish Sea. Series of samples were taken using a 0.1 ma Van Veen grab. Analysis of the amphipods present showed that abundance and the number of species was greater in the west.

One genus-*Bathyporeia-dominated,* but zones dominated by individua species *of Bathypo eia* could be identified also. These could bN correlated with physical variables. Sea water temperature varies in the deeper western Irish Sea from 7 to 13.5°C only, whereas in the shallower eastern areas it ranges from 5 to 16°C.

Research has shown that pollution may depress species diversity in the east and in Liverpool Bay in particular. Sediment size may also be a factor influencing the distribution of amphipod communities.

Bathyporela guilliams niana is the largest species in the genus and is able to inhabit coarser sediments. Sand particles are manipulated by the mouthparts and organic matter scraped off and passed to the mouth.

Bathyporeia nana, dominant in the west, is effectively excluded from coarser sediments because of its small size. Groups of amphipod species often occur together and are good indicators of the shallow-water sand community.

Echinoderms: The heart urchin *Echinocardium* and the little heart urchin *Echinocyamus* live buried in sand. A starfish *Astropecten* also inhabits sand and feeds on small bivalves. The sea cucumber *Holothuria* prefers muddy sediments and captures prey by entwining them with sticky mucous threads. An abyssal sea cucumber.

Benthic Organism—Epibenthos

Organisms living on bottom surfaces (epibenthos) differ

markedly depending on the nature of the substrate.

Hard substrates In shallow cool temperate waters kelp forests dominate and the hold-fasts, stipes and fronds provide suitable surfaces for many fixed and mobile invertebrates to live.

Fixed forms associated with kelp and rock surfaces include many coelenterates such as hydroids, anemones and corals, sea mats, barnacles, mussels attached by byssus threads and sea squirts.

The mobile species include some of the larger decapod crustaceans such as crabs but also many polychaete worms and echinoderms such as the sea urchin *Echinus esculentus.* Under rocks, pebbles and seaweeds large concentrations of brittlestars may occur, e.g. *Amphipholis squamala* which is recorded from lower shore down to 250 m.

One major benthic community is tropical regions that produces its own hard substrate is that of corals and coral reefs. The reefs are formed from the calcareous skeletons of the colonial coelenterates. Distribution is limited to shallow seas where the temperature does not fall below 18.5°C; coral reefs are one of the most biologically productive of all communities.

Water flow over reefs and the efficient re-cycling of nutrients are the factors largely responsible for the high productivity. Although the polyps are animals, the reef is a virtually self-contained ecosystem with its own food web. Algal zooanthellae live in the tissues of the polyps and other algae live in the hard calcareous part of the reef.

These algae need to photosynthe-size and hence restrict live coral to 90,m depth at most. This symbiotic relationship between the coral and algae is thought to have contributed to the success of the reef ecosystem. The relationship appears to allow nitrogen and phosphorus to be recycled.

Hence the typically low levels of these nutrients in tropical waters are no longer a limiting factor in the reef eco-system. The coral polyps also contribute to reef productivity by capturing zooplankton and phytoplankton as the waves pound in. Reefs appears as 'oases' with a gross production greater than 10 g per square metre per day surrounded by open ocean 'desert' of very low productivity.

Some corals do not have symbiotic zooanthellae and are capable of living at greater depths, but they are generally solitary and not reef-forming. Darwin described three types barrier reefs found along continents. fringing reefs around islands and atolls, horseshoe-shaped reefs centred around a lagoon.

Living reefs that are located in mid-ocean are usually situeted on the top of a volcanic cone. In the some atolls there is a great depth of dead coral and this is used as evidence of sea-bed sinking or rising sea levels. The coral has been able to keep pace with the change by growing upward, but any below the photic zone dies.

At night the coral polyps extend their tentacles to capture plankton rich in scarce nutrients such as phosphorus. Vigorous growth occurs in regions where there is good current flow, especially through channels in the reef. On the inner edge of the lagoon silting may depress growth while surf breaking on the outer edge may damage new coral growth; encrusting coralline algae thrive in these conditions of pounding surf.

Reefs attract a dazzling array of other inhabitants ranging from conspicuous shoals of fish to anemones, crustaceans and sponges while nearby, the top predators-sharks and moray eels-lie in wait for the unwary.

Soft Substrates

In deeper waters there is a tendency towards a uniform muddy bottom. Around the poles this is principally composed of diatomaceous ooze, mainly of silica whereas Globigerina ooze dominates in temperature and tropical oceanic regions and is mainly calcareous. Some of the more unusual epibenthic fauna including Venus's Flower Basket Euplectella, a deep-sea glass sponge living between 450 and 900 m.

A commensal relationship has developed with a species of shrimp. A young male and female shrimp enter the hollow cylinder formed by the fusing of supporting spicules into a lattice.

The end plate sealing off the cyiinder effectively imprisons the shrimps which feed on plankton carried on water currents through the lattice. Coelenterates related to the corals known as a sea pens Pennatula have a large primary polyp anchored in the substrate; feeding polyps occur as branches off the main stem.

Priapulid 'worms' are predaceous and capture slow-moving invertebrates by everting the invaginated mouth region armed with rows of spines. A deep water asellote isopod has elongated legs which enables it to move easily over the surface of soft ooze. Other epibenthic fauna photographed from a deep-water trawl.

Melobenthos

This important group of animals includes all those interstitial forms living in the spaces between sand grains. The type of animals range from rotifers, nematodes, harpacticoid copepods, kinorhynches, turbellarian worms through to minute sea cucumbers and sea squirts. Meiofaunal populations of 4 million m^{-1} are not uncommon in shallow seas and their biomass average 10 to 20 per cent that of the macrofauna.

Their importance in benthic ecology and in benthic food webs in particular is only beginning to be appreciated. Many microfaunal groups such as bacteria, protozoa and meiofaunal nematodes have representatives capable of living in deoxygenated or anaerobic conditions.

It has been suggested that the origin of life on earth may have taken place in Precambrian deoxygenated estuarine muds and that early metazoan forms were benthic anaerobes absorbing nutrients through their outer epidermal layer.

Further, it is claimed that anaerobic metabolic pathways were able to switch to aerobic respiration when the oxygen levels in the atmosphere had built up following the development of photosythesis in autotrophs.

Community Analysis

The idea of classing marine benthos into communities was largely established by Petersen in 1913 when he divided up the fauna of Danish waters into communities based on a survey of the animal caught in his grab sampler.

He found that there were certain common species, especially of molluscs and echinoderms, that would act as indicator species. From even a small sample it was possible to predict the kind of community from which the sample had been taken by looking for these indicator species.

He usually chose a pair of species to name a community such as the Echinocardium-filiformis community typical of sandy mud at intermediate depths,

Two of the most conspicuous species are the heart urchin *Echinocardium cordatum* and the brittlestar *Amphiura filiformis* and the com-munity has been named after these indicators.

Other com-unities have been named in a similar way. However, there is still controversy over the best way to classify marine benthic communities. An alternative method favoured by workers such as Jones sampling the benthos off the Isle of Man in the Irish Sea used physical characteristics of the environment, a major division being hard or soft substrates and in the latter group grain size of the sediment.

This led some to suppose that there was little evidence that large assemblagesof animal species interact as a unit or a 'superorganism' but are governed simply by the physical attributes of the environment. A third method of classifying communities includes the type of metabolism used.

This involves looking at the amount of energy tied up in those organisms comprising the community and the flow of energy through the system. The type of community analysis chosen by researchers may depend also on the area in which the work is carried out. If there are sharp physical boundaries then discrete com-munities often coincide with the physical zones.

In areas of gentle gradients, such as often occur in the deep-water benthic habitat, researchers prefer the continum concept with e gradual change in the proportions of species comprising the communities over the area studied.

Henery Bigelow wrote, 'When one picks up a fish one may be said, allegorically, to hold one of the knots is an endless web of netting, of which the countless other knots represent other facts, whether of marine chemistry, physics or geology, or other animals and plants.

And just as one cannot make a fish-net until one has tied all the knots in their proper positions, so one cannot hope to comprehend the web until one can see its internodes in their true relationship.

Clearly, the more one knows of deep-sea food webs the closer one comes to compre-hending the endless web of life. The prospect is daunting, for there are many cross connections between the knots, which are continually being tied and untied. There is immense variation between one generation of knots and the next, from a few hours for a bacterium to many years for a whale.

In the deep ocean our knowledge of food webs is far from adequate. Indeed, we do not even have a proper inventory of the living knots: there are still many new species to be discovered and described.

Even so, systematic and biological investigations are advanced enough to give some understanding of how the food web is made and maintained, though we may well agree that '...the single most important unresolved problem in the fields of marine biology and chemisty concerns the '.. mechanism by which organic matter is trans-ported to depth and how the energy requirements of the deep seas are met'.

Our preceding surveys of oceanic plants and the animals of the pelagic and benthic environments have been partly con-cerned with patterns of productivity and animal means of making a living. It seems clear that benthic animals are largely dependent, whether directly or indirectly, on organic materials' that have fallen from, and through, the food webs of the pelagic zones to eventually reach the bottom. First then, we must look at trophic events in the pelagic sphere.

Chapter 4

CONCEPTS IN MARINE POLLUTION

The *Intetnational Oceanographic Commission* (IOC) for the United Nations Educational and *Scientific Commission* (UNESCO) d.-fin.-s marine pollution as: "Introduction by man, directly or indirectly, of substances into the marine environment (including estuaries) resulting in such *deleterious* effects as harm to living resources, hazards to human health, or hindrance to marine activites and reduction of amenities."

If we allow broad interpretation of the word "*substances*" this definition appears to be all encompassing. Note however the strong contrast between the dictionary and UNESCO definitions. The *connotation* of the *oceanographers* is that unless man *introduces* the *pollutant*, there is on *pollution.*

Apparently the inference is that if pollution is man-caused it can be *mancured.* Nearly 71% of the earth's surface is covered by oceans, which together comprise a total of approximately 1.37 x $10\$^{8}$ litres, weighing about 1.40 × 10^{98} metric or long tons. This is the *tremendous* amount of water and consequently for many years the oceans are regarded an 'ideal place to *dump* all of man's wastes.

For example, the rivers of the world discharge close to 2×10^{14} metric tons of water into the ocean every year but this is much smaller than the water already into the ocean. Carried along with the water each year are 4.5×10^{12}

metric tons of dissolved matereials. Although this is a large amount of material compared to that already present it is literally a drop in the *bucket.*

Unfortunately, the ocean does not receive all of its dissolved material input form the world's rivers. Some comes from the atmosphere as it is washed by falling rain.

Most of the pollutatnts placed into the atmosphere and up in the oceans either directly or indii ectly from precipitation. This also is a sizeable amount of material.

It is to be noted that this is larger than the amount of sediment plus the amount of dissolved materials carried to the oceans by river.

The more important reason for being concerned about marine pollution stems form the fact that materials put into the oceans by man's activities is not evenly spread throughout the oceanic volume, pollution is generated where man lives and works and therefore finds its way for the most part into the oceanic areas that are closest to their activities.

Consequently, the dilution factors is very small and the concentr-ations of these pollutants build up in coastal regions. They are increased to a much higher level than would be the case if the material were distributed throughout the entire oceanic volume.

Coastal are dumping grounds area having a much higher pollution concentration not only because the material is being put into these relatively shallow areas much more rapidly than it is being carried away by natural weve nations but also because the normal structure of the oceans tends to prevent the mixing of these inputs with the rest of the oceanic volume.

The oceans are not dead from pollution, not are they insensitive to the effects of pollutants. As long as we find it necessary to use the sea for a world dump, and there is no reason to believe we will ever stop, we must learn as much as we can about oceanic processes so that the assimilative capacity of the oceans is never exceeded.

Strictly speaking marine pollution may also be of natural origin. For example, it may be caused by underwater gas and oil eruptions, or by the intensive growth and subsequent death of certain micro. organisms, which are responsible for the familiar disastrous 'red tides'.

However, such processes and effects are not regarded as pollutation', even though the are often studied by methods similar or those used for investigaing pollution effects caused by man.

Ecological studies on marine pollution fall into three principal categories

a. Biogeochemistry of pollutants.

b. Marine ecotoxicology.

c. Biological principles of anti-pollution measures.

The biogeochemistry of toxicants includes the investigation of the sources of pollutants, the pathways along which they enter the marine environment, patterns of accumulation in the biotic and abiotic components of ecosystems, mechanisms and rates of migration of pollutants, their transformations and other processes which determine the fate of toxicants in the sea.

An important, though not decisive stage in such investigations is the determination of the content of *toxicants* in marine organisms and communities; this is a difficult analysical problem because the chemical composition of matter is complex, while the *concentration* levels of pollutants are low. The methods of chemical oceanography are used in order to obtain and analyse samples, the results being *interpreted* in a biogeochemical sense.

The second type of research work carried out on this problem. which is best described as '*marine ecotoxicology*', involves the study of biological effects and consequences of human interference with the composition of the marine environment.

Finding answers to such questions, is not by itself tantamount to solving the problem of pollution, but nevertheless yields an objective evaluation of present-day ecological anomalies in the World Ocean and prediction of those likely to occur in the future These investigations provide a scientific foundation on which to base national or international efforts to preserve the seas from pollution.

Solutions of the immediate practical problems concerning pollution of the sea must be based on the results of closely interlinked ecoloxicological and biogeochemical studies.

Marine Pollulants

In table elsewhere in this chapter a list of common pollutants associated with the marine environment have been given. These are not listed in any particular order of importance because it is found that in certain areas certain pollutants are more worrisome than others, while the opposite condition maybe true in other areas.

This difference in importance comes from local conditions like population pressure, flushing, and climate. Theoretical calculations of the concentrations of the principle toxicants in epipelagic waters of the World Ocean indicate that considerable pollution (lead pollution, in particular) of surface waters by atmospheric pollutants may have taken place. This is true of inland seas.

The global pollution zone of the marine environment may be characterised as follows: highest pollution levels in the euphotic layer; high pollution levels in the neritic zone and inland seas; latitudinal distribution of toxic pollutants; a mosaic distribution pattern of toxicant concentrations in water; localisation of toxicants in the hyponeuston and benthos biotope; and a coincidence of the maximum pollution zones with zones of high biomass and high productivity.

Before we discuss the marine pollutants one by one, we shall give some general conclusions about these

(*i*) The rate of biological accumulation and migration of micro-amounts of metals and radionuclides in the World Ocean is directly correlated with their physichcocmical properties and with the specific

Table 4.1: Pollutants associated with the marine environment.

Pathogens	Toxic organics
Sediments	Petroleum
Solid wates	Nutrients
Heat	Radioactive materials
Fresh water	Oxygen demand materials
Brine	Acids and bases
Toxic inorganics	Aesthetically displeasing materials

surface area of equatic organisms.

(*ii*) Concentrations of metals in sea writer and in plankton tend to decrease with increasing atomic number of the metal.

(*iii*) The relative content of artificial radionuclides in oceanic seston does not exceed 1 per cent of their total content in water; the corresponding figure for heavy metals and transition is 1-10 per cent, while in inland seas it may be as high as 25 per cent.

(*iv*) The forms of microcomponents in sea water are largely determined by the supersaturation of surface water by-, calcium carbonate, since the suspended and the colloidal froms of this compound are strong sorbents.

(*v*) Self-purification of surface waters by biosedimentation of impurities is mostly due to their accumulation in aquatic organisms at low trophic levels.

If the accumulation factor in seston is 1000—10000, the rate of vertical biogeochemical migration of a given impurity is of more or less the same order of magnitude as the rate of its hydrological transport (for biomasses and productivities considered as average for the World Ocean).

This is true of many radionuclides (^{144}Ce, ^{95}Zr, ^{106}Ru, °^{5}Zn, b^{4}Mn, etc.), of heavy metals and transition metals present in microamounts (mercury, leads, cadmium, iron, zinc, etc.) and chiro-organic compounds (DDT, PCB, aldrin, lindane, etc.).

(*vi*) The rate of input of petroleum and chloro-organic toxicats into the World Ocean is faster than their rate of decomposition under natural conditions The degradation processes are particularly slow at great depths, and the latter must be considered as a depot of toxic substances and their metabolites.

(*vii*) A typical feature of the geochemical behaviour of petroleum and chlorinated hydrocarbons in the World Ocean is that their distributions in biotic and abiotic components of the same ecosystem are altogether different.

(*viii*) The contemporary radio-ecological status of the World Ocean is distinguished by the relatively stable contents of ^{90}Sr and ^{137Cs} in surface waters and yin aquatic organisms. Another typcial feature is the gradual increase of radionuclide contents in the ecosystems of inland seas.

(*ix*) The concentration levels of artificial radionuclides (^{90}Sr and ^{137}Cs), heavy metals (mercury, lead and cadmium) and chloro-organic compounds (DDT, polychlorinated biphenyl, etc.) in commercially important marine products vary with the degree of pollution of marine biotopes, taxonomic identity of the organisms, and their ecological and physiological characteristics (trophic level in the community, diet, age, zinc, etc.).

The most variable concentrations in industrial—species are typical of substances with cumulative properties, DDT and DDT metabolites, polychlorinated biphenyles and other chloro-organic compounds, and methyl mercury.

(*x*) There is a generally valid relationship (a direct

collelation in the case of radionuclides) between the concentration levels of toxicants in commercial animals and the pollution level of the marine environment as shown by the increasing sequence of pollutant concentrations : pelagic ocean waters, neritic zone, inland seas, freshwater basins.

(*xi*) Typical accumulation factors of ^{90}Sr and ^{137}Cs in commercial species are of the order 10^1—10^2, for heavy metals 10^2—10^8, and for chloro-organic compounds 10^8—10^4.

The food chain effect—i.e., the dependence of the toxicant concentration in a given organism on the trophic level of the organism in the communityis definitely present only in the cases of mercury. polychlorinated biphenyls, DDT and other chloroorganic compounds.

The contents of mercury and of chloroorganic compounds in fish show a regular increase with the size (weight, age) of the animal.

(*xii*) The selective, irreversible accumulation of high concentration of mercury in pelagic predatory fish and marine mammals is a natural process, and is not usually due to human factors.

(*xiii*) The distribution of toxicants in the organs and tissues of commercial fish species is typically non-homogeneous. °OStrontinum, lead and cadmium become preferentially concentrated in ligaments and bones; chloroorganic substances and methyl mercury concentrate in the liver and the lipid fractions; ^{187}caesium concentrates in muscles. Chloro-organic compounds concentrate in fatty tissues.

(*xiv*) Global concentration levels of radioactive and chemical toxicants in commercial marine products are usually much lower than the national and international norms for maximum permissible concentrations, and thus present no danger to human health.

However, in view of the possibility of local pollution and human interference with the natural background of microcomponents in individual regions, concentrations of toxicants in commercial species and marine products must continue to be constantly monitored, especially for heavy metals and chloro-organic compounds.

(*xv*) Generally speaking, monocellular algae my react to the presence of toxic components in the aqueous medium in two different ways

(a) By inhibition of vital processes and production.

(b) By simulation of cell development and cell division, which is usually followed by imapired vital activities.

The development of micro-algae is relatively independent of the presence of the toxicant in the medium.

(*xvi*) The inhibitory effect of the pollutants increases in the following sequence : oil, detergents, metals, (arsenic, lead cadmium, cooper, mercury, methyl mercury), chloro-organic compounds.

The most highly toxic compounds are DDT, polychlorinated biphenyls, mercury and methyl mercury, which inhibit photosynthesis in long-term experiments at concentration as low as 1-10 μe/l. Arsenic, when present as arsenate ion in concentrations of up to 1-10 mg/l, had no effect on the rate of division of algae cells.

(*xvii*) Low concentretions of toxicants (0.1—10μg/l) acting for a short period of time tend to stimulate photosynthesis and the rate of cell division in algae.

This may be caused by the overall activation of vital processes by small doses of poisons or by an alteration of the metalbolic interrelationships with the cells of the microflora. In the case of petroleum products such a stimulation may be due to the presence of biologically active substances in petroleum.

(*xviii*) Toxic (algicidal) and threshold concentration, ranges are determined for various types and species of moncellular algae 'Out of the several combinations of the effects of several factors studied, synergistic effects were observed for the combinations of petroleum with DDT, and combinations of certain metals with a detergent.

A combined effect of several metals may be additive, or else may be manifested by a dominant inhibitory effect of the most toxic one (mercury, copper).

(*xix*) The varying response of different monocultured species to toxic impurities in short-term experiments (less than one day) are probably due to the differing rates of absorption and penetration of toxicants into the cells.

In long-term experiments the specific differences are usually much smaller and depend on molecular, cellular and tissue level mechanisms.

A general characteristic is the better resistance of freshwater protococcal algae to most pollutants, as compared with cultured marine phytoplankton forms.

(*xx*) Of the species studied, Gyrodinium fissum in highly sensitive to mercury, the response being irreversible; the same applies to diatomous algae acted upon by a mixture of DDT and petroleum products.

(*xxi*) The inhibtion of photosynthesis is a more sensitive reactions of monocellular algae to toxicity than is the decrease in the rate of cell division.

(*xxii*) The general picture of the effects and relative toxicities of the most common toxicants on natural communities of marine phytoplankton does not significantly differ from that obtained for cultures of single species.

(*xxiii*) In a number of cases the intensity of photosynthesis of natural phytoplanoton communities decreases in the presence of lower toxicant concentrations than in experiments carried out on monocultures.

The higher sensitivity of natural phytoplankton to contaminats is bably due to interference with the interspecies hips in the community.

The considerable geographic, seasonal and other fluctuations in the reaction of natural phytoplankton to the presence of toxicants are caused not only by the environmental factors, but also by the special features of its specific composition.

This composition undergoes significant changes if the action of the contaminants on the phytoplankton is sufficiently prolonged. Nitzschia diatoms in marine phytocoenoses are more resistant than other species to certain heavy and transition. metals.

Pathogens

The pathogenic materials are the living organisms that can produce sickness or biological unbalance in either plants or animals within the ocean itself or in humans who either come into contact with oceanic waters or eat the organisms caught in the water.

'These include a wide variety of bacteria, protozoa, viruses and fungi. The most common of these are normally found in sewage. However, other pathogens may occur in non-waste disposal areas where environm-ental conditions are such that the proper conditions exist for growth and reproduction.

Sediments

Sediments are always present to some extent in the marine environ-ment. These sediments have a marked effect on plant growth because they block out a large portion of the light normally reaching greater depths and therefore decrease photosynthetic activity. In some case rooted plants get completely destroyed by this process.

As sediments are deposited on the bottom then will cover up bottom dwelling organisms (benthos) such as oysters and may even smother them under extrem conditions. The deposited sediments tend to make the water shallower so that, in regions travered by ships, periodic dredging is needed.

Solid Waste

The disposal of solid waste has been a critical urban problem because areas suitable for the dumping of these valuminous m aerials are becoming scarcer. Due to this reason the ocean has been used as a dumping ground for solid waste.

The solid waste most frequently marine dumped is sludge material left over as a byproduct from domestic sewage treatment. However, the unused products of industry and the used-up products of society are also put into the sea.

If the product discharged at sea contains materials which may be leached into the oceanic environment, a serious problem could arise especially when the materials are toxic. However, when the solid materials are inert, a little forethought may result in some benefit to the marine environment.

Artificial fishing reefs, for example, have been found to be very successful in certain areas, serving to enhance the habitat area, especially for small fish.

Excess Heat

Excess heat, if added to the marine envitonment, alters ambient conditions, and these changes may be detrimental to the organisms present. The amount of heat that is detrimental and the extent of the degradation is determined by a number of factors.

The primary source of this heat has been of course, from electrical generating plants, whether they be fossil fueled or nuclear powered (see Thermal Pollution).

Fresh Water and Brine

Although fresh water may be in great demand ashore, too much of it in the ocean obviously will produce a marked environmental change within a small area. This change may occur due to poorly designed storm drainage systems, water diversion networks associated wish dams, and effluent of some industrial processes. Excessive fresh water usually is not a critical problem.

The introduction of brine into the marine environment is similar in its effect to that of fresh water except in the opposite direction. The organisms acclimated to a particular salinity now find themselves in a more saline environment which could make permanent damage. Brine is a byproduct of desalination, plants.

Toxic

Toxic inorganics have been materials commonly used in industry, many of them are relatively harmless; however, in larger quantities they may be quite destructive. There are perhaps, 35 to 40 commonly used toxic inorganics and these should be very closely controlled by the user.

The toxic organics have been the most disturbing of the modern day chemicals commonly discharged either purposefully or accidently into the marine environment. These include the biocides such as funcigides, herbicides, insecticides, rodenticides, and also the additional organics including halogenated hydrocarbons, petroleum and industrial chemicals.

The most disturbing toxic organics have been the pesticides such a DDT and kepone which have the unfortunate characteristic of being more soluble in oil than in water so that the tend to collect within the fatty tissues of marine organisms. They have been also very stable compounds which do not deteriorate very easily over a long period of time.

These find their way into the ocean both as manufacturing effluent and as runoff after utilization. Their long term effects are unknown upto this fine.

Petroleum

Although petroleum may be classed as a toxic organic, it is a naturally occuring material and is biodegradeable, given enough time. Its effects are not too well known. Petroleum enter into the marine environment due to accident such as tanker damage or transfer loss, natural seepage, off shore production losses, losses associated with refineries, from runoff originating as drippings or disposal of used automobile lubricants, and unburned hydrocarbons emitted into the atmosphere as internal combustion exhaust.

Nutrients

Nutrients (commonly called fertilizers) are those chemicals which are required by plants. The activities of man have addded to the total nutrient load of almost all coastal areas. When nutrient levels becomes out of hand, plants grow unchecked so that decaying plants exist in such great numbers that the oxygen supply becomes rapidly depleted.

These nutrients are present in domestic sewage effluents, agricultural runoff, and the little understood but apparently important nonpoint source runoff from urban areas.

Radioactive

Radioactive material have been not only discharge to the marine environment by nuclear power plants, nuclear power plant fuel production and reprocessing plants, and uranium activities of all sorts, but also result from more common activities of burning of coal.

When coal is burned it gives out radioactive particles to the atmosphere which are the washed into the sea at a greater rate than any known nuclear power plant at this time. Other sources of radioactivity incluce the natural background, weapons testing, mine drainage, accidental spillage, and a few isolated industries.

One of the partially unsolved problems associated with the use of nuclear energy for electrical power has been the long term storage or disposal of spent fuels. This storage must be in an area such there have been no pathways back to man, and the deep oceans have been suggested as meeting this criterion.

Oxygen Demand

Oxygen demand materials have been those which need oxygen for degeneration and therefore steal oxygen which would normally be utilized by marine animals. Thus if too many oxygen demand materials are kept in the marine environment, the animal population will be markedly decreased due to the lack of oxygen.

Sewage sludge and any other organic waste material, even that resulting from excessive plant growth due to an over supply of nutrients have examples of common oxygen demand materials.

Acids and Bases

The discharge of acid and bases to the marine environment is quite disturbing to the natural ec'llogical balance of the system. The normal pH of oceanic water has been somewhere around 8.0, slightly basic. This is maintained by the carbonate system.

If a large amount of acid or base is introduced into the system, the carbonate reactions will be offset and an important element of the environment will get affected. In addition, there are large synergistic effects associated with pH. Most toxic materials, for example, increase their toxicity under conditions of low pH.

The sources of acidic or basic material have been primarily industrial with some of this material reaching the marine environment from accidental discharges and the rupturing of tankers.

Aesthetic Considerations

Aesthetically displeasing material include all the stuff one finds in the ocean which is unpleasant to look at or to smell. Tar balls, floatables, gas (often hydrogen-sulphides) producing materials, and coloring agents are some examples of pollutants offending the senses.

Although in some cases these materals do not pose any real threat to the ecology of an area, when the area is being used for rocreation, the quality of the surroundings becomes some what important.

Management Problems

Although the major pollutants are delineated, the management of marine pollution is quite complex. The reason for that is that the marine environment is used by man in many different ways and this multi-use is having a number of ramifications.

In the first place each oceanic activity pursued by man may produce a whole range of pollutants. A moving ship, for example, may not only discharge pollutants in the form of waste products from the people living aboard, oil from pumping its bilges, and heat from the discharge of its condensers, it may also need deepening of a harbor channel.

This may generate a salinity change in addition to sedimentation problems associated with the maintenance dredging required. Each pollutant may come from many different users. The nutrients which primarily appear to come from the effluent of sewage disposal plants also many come from storm drainage of urban areas in the form of waste products of domestic animals, along with runoff from agricaltural areas that haye been fertilized by either commercial fertilizers or farm animals.

An example of the effect of one pollutant on another has been the attraction of heavy metals to suspended sediments.

Most of the heavy metals in coastal waters are swept out of the water by sinking sedimentary particles and end up in the bottom deposits If the heavy metals then enter their way into bottom dwelling organisms which, in turn, find their way into a food chain pathway to man, then this obviously is not desirable.

Another example of this synergistic effect of two or more pollutants has been the effect of oil on Pesticides, combined with sediments. The pesticides are more soluble in oil than they are in water so that if there is oil present in oceanic water, it will tend to concentrate any pesticides that also may be present.

Oil is also having an affinity for sediments and as it collects on the suspended par titles, their weight gets increased to the point where they sink to the bottom. Portions of the inshore marine environment thus are covered with sediments very rich in both oil and pesticides. The introduction of foreign material into the marine environment has been not always deterimental to man or to the environment either.

In many areas the increased erosion produced by construction activities in large areas of siltation where these eroded sediments have been deposited in the shoreline areas. After stabilization, populations of marsh grass have been seen to spring up in many of these newly created areas.

In the main there is no planning for this type of rehabilitation, but recent experimental work reveals that production of marshlands may very well be a common tool of the esturine planner of the future.

Sources of Marine Pollutants

Now that the pollutants are delineated and described very briefly, we will concern ourselves with the activities involved in the production of these pollutants in an attempt to pinpoint the source of the problem. These activities are outlined in

Table 4.2: Pollution producing activities.
Marine Commerce
Industry
Electrical power generation
Sewage treatment
Other non-industrial
Recreation
Construction

Let us discuss their one by one.

Marine Commerce

Marine commerce has been deeply involved with ships, and ships require a deep enough water environment .Lso that they can pass to form on their appointed rounds.

This usually requires continuous dredging, Dredging is an expensive proposition requiring large expenditures of money. Once the sediments are extracted from the bottom and the channel is cut to the desired depth, the problem becomes one of disposing of the dredge spoils, the material dredged from the bottom.

Various solutions have been suggested but the problem is still one that has not been adequately solved. Another aspect of the ship problem has been oil. Whether these ships has been transporting oil or using it for fuel and lubrication, there always seems to be some lost to the sea.

Most harbors are having large storage areas open to the atmosphere so that a certain amount of leaching into the marine environment is inevitable. Lastly, there has been a whole family of pollutants associated with any ship. A ship is just a little city afloat and any problems found in urban areas will be found on ships also.

Industry

Whenever one mentions pollution to the average individual the first source of pollution that pops into mind has been industry. Most industries require the use of water. This water has either heat or some chemical added to it before being discharged into the environment.

These chemicals may be anything from toxic substances to harmless coloring agents. Electrical power generation is responsible for waste heat and certainly this is one of the by products of producing electrical power whether in be from the burning of fossel or nuclear fuels.

An additional problem involved with heat exchangers has been the fact that many of these are made from copper, a toxic material which is used in anti-fouling paints on boats for many years. Sometimes these heat exchangers are allowed to remain in an unused condition for extended periods of time with water sitting in the heat exchanger tubes.

Some copper leaches into this water from the tubes and when the plant gets turned on again this copper-rich water is pumped into the environment where it may bring about some consternation among the local marine organisms.

Large amounts of water pumped at high speeds from a coastal regions are able to trap or entrain small of sometimes even large organisms. These creatures may suffer mechanical damge as they go through the system, or they may get exposed to large amounts of beat for a short period of time.

Radioactive wastes are considered as one possible disadvantage of nuclear power plants, but the amount of radioactive wast actually transferred by cooling water to the surrounding environment has been measured as extremely low. The spent fuel problem appeprs to be of more concern at this time.

The pollution problems associated with mineral extraction could not be documented due to the small opportunity for data taking, but it is expected they will be similar to marine construction. At this time the major oceanic mining effort has been directed toward the extraction of sand and graved in the near shore areas. Within the imminent future it is expected that a connected, effort will be made to extract manganese nodules frog the ocean floor, perhaps at depth as great as 3000 metres. This mining effort will be expected to have some effect on the environment but just how much has been not quite clear.

The same can be said for the increase in activity connected with offshore oil drilling although it appears that there is less petroleum lost into the environment when the oil is obtained from an offshore well and transported to the continent by pipeline than when it is obtained from a land well and transported by means of tanker to its eventual user point.

The spillage resulting from the use of tankers has been greater than that resulting from the use of pine line.

Sewage Treatment Systems

The process treating human waste products has been quite successful in removing suggested materials and pathogens, but it has been extremely difficult to extract all of the nitrogen and phosphorous compounds.

Consequently, sewage treatment facility effluent is having reasonably high amounts of these nutrients which push plants toward uncontrolled growth. In addition, when the facility gets overloaded as during a heavy rain, it is often completely bypassed, discharging raw sewage directly into the water.

This needs a great deal of oxygen for decomposition so that the biological oxygen demand is increased.

One aspect of sewage treatment neglected up to now, primarily because the costs have been so overwhelming, has been the treatment of storm runoff. Unfortunately the solution is not as straight-forward as simply increasing the capacity of existing plants to handle the heavy loads associated with storms, or even to add treatment to storm drainage systems when they exist separately.

Much of the urban and rural runoff does not go through conventional pipe systems, so additional systems would have to be built if all runoff has to be treated. Storm drainage washes urban streets clean. All material deposited on the streets, including crankease drippings, rubber worm from auto tires, and bird, dog, and cat droppings are washed to sea.

Modern studies have indicated that this 1 unoff has been a very significant source of califorms and petroleum, to name just two pollutants of some concern to man. How to handle these non-point sources has been a task presently facing marine environment managers and one that appears to have to inexpensive or simple resolution.

With most large urban areas disposal of the solid residue from sewage treatment plants has been a solid waste disposal problem of the first magnitude. Ocean dumping in the nearshore coastal regime is utilized for many years and, to a certain extent, will probably continue to be.

Either the sludge must get treated more thoroughly before it is dumped or new deep sea dumping means must be found. If new treatment or disposal techniques are not followed, there may occur major changes in the coastal environment.

Agricultural Sources

The largest single class of non-industrial polluters are farms. Agricultural pollutants have been commercial fertilizers, animal wastes pesticides, herbicides, and sediments. In many cases the agricultural contribution of all these pollutants has been greater than any other single source.

The polintion can be stopped, and there have been cases in the past where particularly had problems have been solved.

Recreational Activities

Usually when we think of marine recreation we think of a non-polluted area since recreation is essentially limited by pollution. In shallows water motor boats stir up the sediment on the bottom, a condition which often persists for long periods of time.

This increase in turbidity markedly influences the natural plant growth and probably has something to do with the decrease in the numbers of rooted plants reported in the last few year. The aspect of motor boating that has many investigators most concerned has been the discharge of petroleum products into the environment.

There are some figures which indicate that recreational boats add as much petroleum and petroleum waste products to the environment as all other source combined.

CONSTRUCTION

The last source of pollutants listed in Table elsewhere

in this chapter has been construction. By construction has been meant both that done at sea such as offshore oil rigs, and than done on land contiguous with the sea, such as port facilities. Implied here has been both the actual construction of the structure and the associated activities that go along with construction. There must he all sorts of support equipment, but most important, there must be support people.

As in general, the amount of pollution get directly related to numbers of people, any large construction effort will bring with it the usual peoplewaste products. The effect of a new physical structure where one had not previously has been not too well known, but one thing is sure the environment will never be the same as it was before.

Lome of these changes might be beneficial, but no matter how beneficial there have been aways going to be some members of society who react to any change by claiming a degradation of the environment. Any marine construction project must get judged on the basis of the benefits to be derived by society as a whole rather than solely to those accruing to some small group.

EFFECTS OF MARINE POLLUTION

In this article, we shall attempt to get some idea as to what pollution does and why we consider it to be so undesirable. There have been three basic types of pollutants; the pathogenic, the aesthetic, and the ecomorphic. Pathogenic pollutants have been those which cause disease. This disease may be fatal if the pollutant is a fethal poison.

Aesthetic pollutants have been those pollutants causing a change in the environment displeasing to the eye, ear, or nose of man. Ecomorphic pollutants, on the other hand, have been those pollutants which produce a change in the physical characteristics of the environmeat in such a way that there may be drastic changes in the structure or composition of the biosphere.

Obvious!y, these three types of pollutants have been

not of equal importance, Pathogenic pollutants have been certainly much more serious than the other two : however, e.omorphic pollutants are often an indicate of more serious types of pollution to follow. Furthermore, since pollution has been man-product and the effects have been suffered by man, aeshhetic problems are certainly of interest.

Acute Effects

The pathogenic pollutants have been obviously the most important. The effects of pathogenic pollutants are either acute or chronic.

The acute effects have been the easiest to determine because of the short time between administration and affliction. The easiest way of measuring acute effect has been by feeding the pollutant in question to test organisms, such as mice or selected fish, and observing the dosage required to kill 50% of the organisms involved.

This. is usually carried out by feeding a population of test organisms increasing amounts of the pollutant, and for each dosage the number of individuals that succumb is noted.

When these data are plotted, an S-shaped curve results. Another way of examining acute effect is to observe the effect of a given pollutant concentration on organisms over a somewhat longer period of time, noting have long it takes for a given dosage to produce mortalities.

This may be done by giving a predetermined concentration of the pollutant to the test animals and observing how long it takes to kill half of them. When data from an experiment such as this are plotted, a curve similar to vertical asymptote results.

Chronic Effects

There appears to be a dose size for each pathogen below which the pollutant does not have any effect. The concern with threshold values for various pollutants has been one aspect of the study of chronic effects.

Chronic effects are the most difficult to measure because

of the time involved; in some cases concerns have developed as long as 25 years after iniri it exposure to the carcinogenic agent. Nevertheless, measurements are made and for many materials a thi eshold value seems to describe the data reasonably well.

Most experiments require some measure of the effect on the organism other than death, so that mans different measures are used. These are; deformity in the growth of organisms, damage to particular organs, change in heartbeat or breathing rate, genetic damage to individuals (rather than to the population as a whole), change in the rate of increase of a population, life expectancy of individuals within a particular population, and fecundity of females within the population. Unfortunately, an index may be very descriptive of a pollutant on a particular species, but it may be completely inffective in describing what happenes when this pollutant comes in contact with other species.

Synergism

Another complication in the examination and quantification of pathogenic pollutants has been the phenomenon of synergism. A synergistic effect takes place when the combined effect of two or more materials acting together has been gaeater than would be expected from the simple sum of the individual effects.

Temperature, for example, has been a very strong synergistic parameter, generally enhancing most pollutant effects as temperature is increased.

The same has been true of salinity and dissolved oxygen. Some pollutants, on the other hand, tend to decrease the effects of others, as might be in the case of strong acid and a strong base present at the same time.

Other pollutants may make strong pathogens to get precipitated out of the water column or to be collected on sediments, so that in either case they end up on the bottom Thus it has been extremely important in the analysis of any pollutant to be aware of the other materials present and their synergistic tendencies.

Within the marine biosphere many organisms get effected by various pollutants. Whether or not a particular organism has been effected by any given pollutant will depend on many variables.

For examples, many pollutants occur only in the water column while others, such as some of the pesticides and heavy metals, are adsorbed into suspended sediments and usually find their way to the bottom within a relatively short period of time.

Consequently, these latter types could tend to affect those organisms living on the bottom or feeding from bottom organisms more than they tiould infuence the organisms living within the water column, like free swimming fish.

Also of importanee in determining the effectiveness of a particular pollutant on living organisms has been its relative solubility in water and oil. Pollutants more soluble in oil will tend to find their way to organisms that have a larger content of oil in their body tissues.

This has been particularly true of some of the pesticides which are more soluble in oil than water. The individual habits of marine organisms also influence their susceptibility to a particular pollutant.

Feeding habits with respect to time of day, portion of water column foraged, type of material ingested, and the method of digestion utilized all determine to a great extent the types of pollutants the organism will be exposed to.

Similarly spawning habits decide the type of area and time of year in which a species has been particularly sensitive to subtle changes in the environment. These changes influence not only the parents but also the offspring, and may be even their progeny.

Other habits not having to do with spawning also help to determine whether an organism will get exposed to particular pollutants.

Whether the organism has been sensitive to light or sound, for example, will often determine whether it frets attracted to a particular out-fall. Thus we find that a pure and simple cause and effect relationship between a pollutant and an organism has been not enough to determine completely the actual response of the organism in the real world.

Pathways to Man

Once some of the pollutant gets ingested by marine organisms, the major concern of man has been possibility of this pollutant appearing on his dinner table and causing a pathogenic response. Obviously if a fish ingests a pollutant and a human eats that fish, then that person will get ingested a dose of that pollutant and these will probably be some danger. The question **is,** "How much ?"

In order to arrive at an approximation of the actual amount of undesired material finding its way into the reader's stomach, one must be able to follow the pollutant through a number of steps. Most organisms living in the sea tend to concentrate materials existing in the sea to much greater values than they have been in the ocean itself.

For example, a diatom, a small form of marine algae having a silicate frustule, is a having a body concentration of silicon about 40 thousand times as great as the oceanic waters from which the organism derive; all of its material.

Thus the diatom concentrates silicon very effectively so that if we are diatoms, a large portion of our diet would be glassy. Similarly, other organisms, including fish, will concentrate other materials even if these do not naturally occur in the evironment. But this implification or concentration does not stop with one step.

We must continually keep in mind the fact that an organism does not exist by itself in the marine environment. Any organism has been dependent upon many other organisms for its existence. An edible fish, for example, might very well subsist on smaller fish which, in turn, might subsist on small zooplankton (small floating forms of animal life), which, in turn, may subsist on phytoplankton (forms of floating plant life).

There may be 5 to 10 individual links in a food chain leading to man, and there has been the distinct possibility that for each one of these links a concentration of the pollutant occurs. Hence, the total pathway to man for the pollutant might be a ralher tortuous one, but it might result in a very high concentration of the pollutant material in the fish.

This in itself may be of no concern to the individual unless he ingests the fish and if so, how often and how much he consumes. It has been not enough simply to know that there has been some pollutant present in marine organisms; one should also know how much is present, and in addition two other facts have been required.

This first has been the accepted maximum level of ingestion of the pollutant before harm results, and the second has been the amount of fish that can be eaten before this level gets reached. In this way it might very well be possible to have a particular kind of fish one or two times a year with perfect safety, while if it has been eaten once a day, it might very well have a toxic effect.

Ecomorpbic Pollutants

There are many different types of pollutants which may be classified as ecomorphic, while some pollutants fall into all three of the classes stipulated above and many fall into at least two.

Sediment has been a pollutant whose effect has been primarily that of changing the environment since it is not pathogenic and very often is not aesthetically displeasing, although it can be. As it settles to the bottom, it will tend to bury the organisms thai live on the bottom, such as oysters and clams, and eventually smother them.

Sediment also tends to fill in marsh areas, killing the marsh grass, and completely changing the habitat. As marsh areas are breeding grounds for small fish and an important ecological link in most of the oceanic life, any destructive process such as this has been bound to markedly change the marine biological environment.

On the other hand, in some cases where man's activities are able to produce large amounts of erosion and sediment has been brought down by rivers into estuaries, some estuarine areas have been filled enough to support marsh grasses. In same cases excess erosion can produce marsh lands rather than destroy them.

Whenever there has been a depth change in an estuarine or coastal area, the effects have been many. Tney range from changing the biological environment so that larger fish will no longer live in the shoal areas to affecting the total circulation pattern of the area.

If the sediment has been in suspension it also has an effect on the ecosystem, the most obvious aspect being a change in transparency. By making the water more opaque, light will not penetrate as deeply as before so that the thickness of the layer of water in which plants grow (the euphotic zone) will get decreased.

If the water has been fairly shallow to begin with, this decrease in transparency may very well result in the inability of rooted plants to grow and some marsh lands may be destroyed. The effect of sediment, then, has been two-fold; changing both bottom and water properties. Both of these effects tend to change the physical characteristics of the ecosystem.

Another ecomorphic pollutant has been oil, although oil sometimes produces toxic effects. The result of oil coating bird feathers has been a well known physical effect. The oil does not poison birds; it simply makes it impossible for them to fly, and consequently many of them die.

A film of oil on the surface will also decrease the amount of sunlight entering the water and, similarly, will limit the exchange of oxygen from the atmosphere. In addition to the aesthetic and pathogenic effects of oil, there are thus also ecomorphic effects.

The toxic effects of various nitrogen compounds have been well known but fertilizers and other nutrients washed into the sea possess a nontoxic effect on marine waters.

These materials will do just what they have been designed to do: make plants grow, whether these plants be terrestial or aquatic plants.

Consequently, the first effect of a spike of fertilizer added to the sea has been an overstimulation of those plants normally there. Beyond that the undesirable plants (those that are not food for the edible types of marine animals) often get stimulated grow and these marine weeds grow so rapidly that the overwhelm the other types of plants.

As plants have been at the bottom of the food pyramid, the structure and characteristics of the ecosystem have been completely changed, even to the extent of markedly altering the species composition.

Once these plants have started to grow at greatly accelerated rates, they completely cover the surface of the water, disallowing any light from penetrating more than a few centimetres.

Those plants on the bottom of the blanket will die, due to lack of light, and those rotting plants will use oxygen in their decay processes. Even though there have been more plant than before, producing more oxygen in the process of photosynthesis, so many of them are dying and decaying the result has been co decrease the dissolved oxygen.

This phenomenon is termed as eutrophicatgion. Solid waste has been another type of pollutant very often disposed of in oceanic areas. In some cases they tend to completely destroy the natural habitat and the result has been not only ecomorphic but also aesthetically displeasing.

However, in soma cases, the ocean dumping of waste products such as old automobile tires and even automobile bodies has produced an artificial for many organisms that were not endemic to the region previously.

However, there are cases where nontoxic solid waste materials have destroyed desirable habitats. Sometimes the dumping of solid wastes makes a change in water flow patterns producing changes in salinity and temperatue which influence the biosphere markedly.

This usually takes place when the water is reasonably shallow and the volume of solid waste products dumped is relatively large. Anoter ecomorphic pollutant that seems to be receiving a lot of altention lately is heat which is supplied to the marine environment from electrical power plants or other industrial sources.

This heat will increase the water temperature and may result in different kinds of damage to marine organisms. As most mutations have been caused by thermal activity under normal conditions, the possibility of genetic damage has been real.

It is estimated that about 90% of the mutations occuring mormally have been thermally caused, while only about 10% have been caused by radiation. Thus with an increas in the temperuture, one can expect greater mutation rates or more genetic damage.

In addition to genetic damage, heat can cause both acute and chronic problems Animals can be killed by extreme thermal shock or they can suffer lingering effects and finally expire by being unable to acclimate to the new surroundings.

More subtle changes are also noticed as growth rates are changed with increase or decrease in temperature. Normally a small increase in temperature will make a small increase in growth rate; however, if the temperature change in extreme, this may be reversed.

Changes in spawning patterns have also been noted in which fish that normally move upstream to spawn are unable or unwilling to cross a thermal barrier associated with a heated effluent. This may cause the cessation of spawning for a large portion of the species population.

Another effect of temperature increase has been to cause certain organisms to grow more rapidly than others upsetting the previous ecological balance and eventually causing a change in species composition. A similar range of effects to that produced by heat has been caused by salinity changes.

These might be produced from the input of rich brine solutions resulting from the effluent of warer desalination efforts or a fresh water outfall such as a raw sewage treatment plant A marked change in salinity will bring about damage similar to that produced by heat, including genetic damage growth rate changes, spawning pattcrn changes, and changes in species composition. All organisms are used to living within some range of salinity and if this range gets exceeded, they will either die or move some place else.

Aesthetic Pollutants

The last type of pollutant mentioned above was aesthetic pollutants. Aesthetic pollutants have been those which offened the human senses and as such are regarded by many to be unimportant. In many environmental problem areas, though, the aesthetic considerations often receive more attention by environmentally oriented citizens than some of the pathogenic pollutants.

This is simply because the aesthetic pollutants have been visible and what can be seen, smelled, or heard appears much more bothersome than that which is hidden, even though the hidden pollutant might be much more dangerous to life.

Consequently, any scheme for controlling marine pollution must include some consideration of the aesthetic pollutants if it is to succeed. An excellent strategy seems to be to focus on the aesthetic pollutants and use them as a mechanism to gather momentum to clean up the rest of the pollution problems.

Solid waste material, when piled in shallow areas where it can be seen, has been a typical aesthetic pollutant. Building activities also may be regarded to be sources of aesthetic pollutants. Power transmission cable has caused more than one power company grief because of the environmental consideration given the local citizens who are concerned about the appearance of power cables running across the countryside.

Oil wells, especially the offshore variety visible from resort areas, have been of great concern to most environmentalists. The same thing has been true of equipment required for dredging operations and the spoil areas resulting from these operations.

To some people, especially those in living relatively pristine environments near the water, boats can also be an aesthetic pollutant, especially those that cause waves and noise. Effluents that make the color of the sea to change or increase the amount of suspended sediments also may be regarded aesthetic pollutants.

It simply has been not as pleasant to go swimming in water that is cloudy or contains coloring material different from what one would expect in a normal, natural body of water.

In essence, any change from the pristine environment appears to many to be aesthetic pollution. Obviously, we are living in an unreal world if we think the marine environment can be maintained in an unspoiled form. It should be kept in mind that many complaints about aesthetic pollution come from individual who would rather live in the primeval environment not realizing that it would not be available to them unless the commercial world about them supported it.

There appears to be a dichotomy wherein we try to maintain the unspoiled environment in certain areas while letting others become overutilized. Neverthless, there has been a growing movement to include in the cost of every project a relatively small sum to improve its aesthetic qualities.

This is, of course, exactly the same philosophy that has been applied to the control of other forms of pollution. Additional money has been spent in a chemical process, for example, to control the pollutionproducing properties of the effluent because society believes that this additional sum of money has been a desirable expenditure.

It may be similarly be maintained that the expenc iture

for aesthetic pollution control is desirable as it also will tend to increase the quality of life.

CONTROL OF MARINE POLLUTION

Now that the nature of pollution and some of the sources of this undesirable product of society have been considered, we will now concern ourselves with some of the avenues open to us to accomplish some sort of control.

The most obvious control strategy that can be used has been to some-how or other completely prevent the introduction of the pollutant into the environment. This can be carried out by not producing the pollutant or by restructuring the process so that the pollutant is no longer a by product.

Let us consider a few major sources of pollutants and some possible methods of control in these particular cases.

Dredging

Maintenance dredging, that is, dredging required to maintain a particular depth, adds sediment to the water column and would not be needed if there were no sediment supply to shoal up major channel areas.

This type of dredging has been not a one shot affair, but it has been almost always necessary to periodically redredge the occumulated sediment from these channels and harbor areas. This continual sedimentation has been a normal process which occurs whether man is present or not. In some .stances the activities of man tend to accelerate the sedimentation process but this has been not necessarily always true. For example, coastal plain estuaries, often the result of river basin flooding, tend to be a rather ephemeral event in geologic time.

They get formed when water level rises, and they disappear with the continual process of deposition of material eroded from the highter reaches of the water shed.

Another example of natural erosion has been found on the ' west coast of the United States in the westernmost portion of California (that portion where more than 90% of the population lives). This area was formed by erosion of the Rocky Mountains and deposition of the eroded material at the base of the mountains.

The most obvious solution to the problem of maintence dredging has been to cut off the sediment supply to the areas where sedimentation has been a major problem. This can be carried out in a number of different ways.

The most effective of all has been to prevent upstream erosion, but at the same time this is perhaps the most difficult. The entire watershed must be policed very carefully to make sure that the natural landscape gets changed 'o such an extent that erosin no longer takes place. This implies extensive terracing and replanting of areas that have been denuded for one reason or another.

Of course, this also implies that extreme measures must be taken in areas that do not have natural cover, such as agricultural areas or areas where construction is underway. A certain amount of this is being done.

Another method of pi eventing sediment from reaching the undesired are has been to provide settling ponds. The water is released after it has dropped the major portion of its load and is allowed to proceed downstream.

The major disadvantage of this system has been that it is essentially a temporary solution since the pcnds will fill up rather rapidly, needing either building of new ponds or dredging of some sort. Sometimes the sedimentation that takes place in a channel area is a result of sediment loads that are introduced into the marine environment from nearby rather than those carried from far up-stream.

This takes place for example, when a channel gets dredged with unstable sides so that the material just tends to slough into the deeper regions. This needs continual dredging as the removed sediment has been almost immediately replaced by bottom material from the sides. One method of correcting this situation is by stabilization of the bottom. Adding larger, more dense material such as gravel to the bottom has been a technique utilized occasionally, although this has been usually more expensive than dredging so that it has been not done too often.

Bottom stabilization, however, is a technique that is being used more and m ore to prevent the movement of bottom materials within relatively small contiguous areas.

Another possibility for preventing the accumulation of sediments in undesirable areas has been to change the natural circulation patterns so that so that sediments get deposited in an area where they do not damage.

This may be carried out in various ways: by changing the physical dimensions of estuarine areas, by putting dams in selected areas, and even by changing the circulation pattern by planting grasses to increase bottom friction.

Another method be considered in decreasing the amount of suspended sediments has been that involved in controlling the numbers and types or water craft allowed in particular areas. In regions that have been retatively shall,)w and narrow, high speed vessels tend to churn up the bottom, putting sediments back into the water column that had previously settled out.

In addition, large wakes are generally produced that tend to erode the sides of the waterway by the mechanical action of these relatively high energy waves breaking on the shoreline. Both these effects ckn get decreased by the simple expedient of limiting the speed of boats in certain areas.

It has been a fairly effective method; however, if the traffic is extremely heavy it may be necessary to limit the total number of vessels, in addition to their speed. In addition to the general clarity aspect of pollution caused by dredging, there has been also a toxic one.

Sediment retrieved from harbor bottoms has been especially likely to contain a relatively high concentration of industrial pollutants such as oil and heavy metals. During the dredging operation there has been a likelihood of some of this material's finding its way back into the water column, but of more concern has been the longtime leaching from dredge spoil dumping areas.

Various methods, such as diking and the construction of artificial islands, are utilized to minimize this leaching but their performance under all conditions has been not too well known. Further research in the handing of dredge spoil gets needed at this time. Another problem that, to a certain extent, appears amenable to prevention at the source is that involving oil.

Unforunately, the sources of oil in the marine environment have been extremely difficult to delineate. All the experts agree that the greatest terrestial source of petroleum products in the ocean is used crankcase off. Some go so far as to indicate that this source supplies more petroleum to the sea then *any* other source.

This may be true because within the last decade or so the habits of American motorists are changed markedly to the extent that a large portion of them now change their own oil, discarding the used oil in a manner that probably makes it to end up in storm sewers.

The major reason for this probably has been the almost complete disappearance of the used lubricating oil re-refining industry in recent years since it was not economically !feasible to re-refine oil. However, with the modern continual increase in the price of new oil, the re-refining industry has been having a rebirth so that a large portion of the oil previously thrown away will now a reused.

Another entirely different source of oil in the marine environment has been that resulting from accidents that take place during shipping and transferring processes. Generally speaking, the more handling steps involved, the more accidents and associated losses can get expected.

The figures as to how much has been actually lost in handling accidents compared to that lost in normal ship procedures, such that as pumping bilges, has been not well known, but there has been no doubt that a very large portion of the oil reaching the marine environment has been the result of poor handling procedures in both the operation of vessels and the transfer of oil from one carrier to another.

Better procedures are required, but perhaps even more than better procedures, stricter enforcement of accepted safe procedures would go a long way to decreasing the amount of oil that enters the ocean each other.

When a tanker empties its load of oil, the dynamics characteristics of this vessel get markedly changed. Its weight gets decreased about a thousaad times and therefore it becomes extremely difficult to drive the ship, because it was designed to be driven with a *full load.*

Therefore, when most tankers do not have a load of oil, they will fill their tanks with water for ballast. When they return for a fresh cargo of oil, they must pump this water from the tanks, releasing large amounts of oil as the residue left in the tanks gets washed out.

There are many schemes which are suggested for decreasing the amount of oil that is wasted in the ballasting-unballasting process; however, it is not clear at this time how effective these processes have been or how effectively and requirements can be enforced.

Modern supertankers have been extremely large vessels and therefore have major structural weakneses, so that occasionally they will breakup in unusually heavy seas. They have been simply not designed to be utilized under unusual storm conditions, so that there will continne to be losses of this type in rough weather.

In passing, it should be metioned that once oil gets spilled at sea under large wave conditions (usually) the case when accidents occur, there has been no way for this oil to be retrieved. On the other hand, oil may be retrieved in harbor areas where the sea surface is relatively calm, but only when there are waves no higher than approximately 1 metre.

The obvious way the only way-to prevent oil from getting to the marine environment when a ship founders has been to prevent the accident from occuring in the first place. Alternatively, once the accident has taken place the oil must be prevented from escaping from the ship. It has been somewhat doubtful whether these solutions will ever be possible.

Industrial Pollutants

Most industries are using large amounts of water for cooling purposes or as an integral part of the manufacturing process. Consequently, industrial effluents may be having waste products, heat, leached material from heat exchangers, or even incidental house-cleaning wastes.

If the industry, for example, has been using water as a coolant, the heat exchangers must constantly be cleaned to keep the heat exchange efficiency high. These cleaners may be more toxic than process wastes. Consequently, it has been extremely desirable to treat all industrial effluents. Some industries have tried to modify the entire process rather than just clean up the effluent.

This often causes an increase in efficieny with a lower overall production cost. If this expenditure can be recouped in a relatively short period of time, this obviously might be expected to be a very costeffective method of htdling unwanted effluents.

However, in many cases, there has been no increase in operating efficieney, with the net result being an increase in manufacturing costs passed on to the consumer. Society must then make a decision as to how much it has been willing to pay for a particular product *vis a vis* the amount of pollution introduced in the manufacture of this product.

Antifouling Paint

Sometimes pollution comes from the use of antifouling paint on boats. During the summer time in temperature waters the growth rate of marine organisms get increased marktedly, especially those organisms setting on fixed surfaces.

An untreated boat hull left in the water for a period of

just a few weeks will need enough growth to cut the available speed for a given power input by fifty per cent. Consequently, all owners of boats left in the water have been very careful to apply some coating to the underside to prevent the accumulation of marine organisms.

For Jar large cargo ships or tankers, this could make savings of thousands of dollars in ful costs. Common practice has been to paint the bottom of the boat with antifauling paint.

Antifouling paint has been so used to leach a toxic material into the water in just the proper amounts so that all organisms setting on the hull will get killed. However, the older types of antifouling paints, those that use copper as a toxin, released enough material to the environment not only to kill possible fouling organisms but also to be harmful to other marine creatures at some distance from the boat.

Thus there has been a movement of to improve the quality of antifouling pains by a drastically reducing this leaching to to toxic material into the water. The solution that has been suggested has been to use new paint vehicle, in conjunction with new toxic materials so that the toxic material has been almost completely retained within the vehicle, requiring organisms to come within a very short distance of the boat to be affected.

One result of recent resoarch is to develop a family of paints using acrylic resins as vehicles and tri-butal tin flouride (TBTF) as the primary antifouling ingredient. In the first few years of use this paint research because it is less toxic in the water, lasts loger, and does not allow bottom growth more effectively.

In their very effectiveness these paints pose a problem, because the toxin gets retained within the paint for a longer period of time. Paint scrapings have been now more toxic than they were with the old copper bottom paint.

This means that it has been somewhat more dangerous to strip the paint from the bottom of the boat when applying a new coat and precautions must be used to prevent the old paint from getting back into the water environment. When sandblasting large ships, for example, precautions have to be taken to retain the sandblasted material rather than allowing it to be simply washed away. Thus the solution of one problem presents another, but these problems have been soluble if some care has been taken to see that additional health hazards do not get created.

Pesticide and Herbicide Pollution

Some of the juicier pollution scandals that have taken place in recent years have involved various pesticides and the problem has been that many of the chemical compounds get introduced for public sale before the total effects of these materials have been known.

It seems to be almost an insurmountable problem to determine what the total effects of a new compound will be before it has been put on the market because so many new products are introduced every year.

However, a large portion of the problem has been undoubtedly due to basic carelessness. In both the manufacturing and use of these materials proper precautions are not being taken into consideration.

Many users, for example, undoubtedly believe that if a certain amount per unit area will suffice, twice as much will produce twice as much control of either weeds or bugs. This excess toxin is washed into the streams and ends up killing marine organisms rather than protecting plants.

It would be evident that more effective control of both the manufacture and use of pesticides and herbicides would go a long way toward eliminating undesirable side effects.

Solid Waste Pollution

The disposal of solid waste has been very necessary and there even seems to be a fear that if the solid water problem is not solved, eventually there will be no room left for people. Because of this fear, large amount of solid waste get dumbed in the ocean.

Some of these solid wastes take the form of sludge from sewage treatment plants whereas others consist of large pieces of discarded consumer products. In the latter case the problem might very well solve itself as the cost of scrap metal continues to increase.

One possible use for sewage treatment sludge has been in refurbishing land disfigured by strip mining. Often the overburden removed to get at the coal seam has been not capable of supporting life nor it has been even particularly stable.

The addition of sludge from sewage treatment plants is considered as a method of roil treatment to make it possible to support a viable plant population. It is a concept that has not been tried yet but one which will be tried when it becomes economically feasible to transport the sludge to the strip mine.

Another possibility for using the sewage sludge, if it contains only nutrient materials, has been suggested and has been considered by some of the Middle Eastern oil producing countries. The idea has been that empty tankers be filled with sewage sludge for the return voyage rather than with ballast water.

This sludge will then be utilized to bring the desert regions back to a point where they have been capable of supporting a large agricultural industry. Other forms of solid waste being dumped in nearshore areas have been dumped there primarily because there has been no other acceptable depository.

One method of decreasing the severity of this problem has been simply to decrease the amount of solid waste produced by society. Packaging of consumer items, for example, seems to be a good place to start. From a disposal point of view the catsup bottle is probably the worst possible package.

The package weighs more and takes up more space than the contents. The Navy has done excellent work in decreasing package size and weight because there are definite limitations aboard ship, but too much has been done with the private sector of society simply because there 'asn't been enough action on the part of consumers to insist upon small volume aad weight packages.

One possible exception to this has been the introduction of lightweight plastic bottles for large size drinks to replace the heavy weight glass variety. They are somewhat thinner than glass, weigh less, and may be incinerated, so the disposal problem has been somewhat alleviated. However, with the continuing increase in petroleum costs, plastics might not remain economically feasible in the future.

Recycling and Reclamation

It has been found that biological communities living in the sea need a certain amount of physical shelter. Some species just like to be alone, whereas in other cases the young of the species require places to hitde to protect against predation.

A marsh or wetlands area has been a good breeding place because both of these conditions prevail. It is found that in many cases these conditions can be created artificially by intoroducing large junk articles such as old automobiles or worn out ships to form effective artificial fishing reefs.

There has been disadvantages to using waste products for this sort of activity. One has been that in many cases these junk, objects get transported to undesirable areas by dynamic oceanic forces. Pollution has been not controlled when automobile fenders and doors wash up on swimming beaches.

When creating a finishing reef, steps must be taken to retain the integrity of these reefs for a long period of time. If properattention has been paid to the magnitude and nature of oceanic forces, artificial reefs night he made relatively permanent.

With the present state of the economy where things are some what more expensive, it appears that more items will find themselves in the near future on the list, of materials to be recycled or reclaimed. Industrial processess will alter as it becomes more feasible to extract materials from the waste stream from a cost standpoint and, at the same time industrial.

Waste streams will become less and less polluted. Even in the area of multiple waste product use, like the combination of waste heat from either manufacturing or electrical power generation and waste nutrients from sewage treatment plants for the purposes of aquaculture, cost will always be the controlling factor.

It is known that elevated temperatures and the artificial addition of nutrients will increase growth rates of many organisms markedly. The choice of which organism to grow has been involved not only with which would be best to grow, but also which there has been a market for.

Setting upon aquaculture system involves picking organisms which respond favourably to the environment, developing a system by which the optimum growth rate can get produced, developing a marketing system so that the end product will be sold at the maximum possible price, and assuring the continuing existence of a market for everything produced.

Not only has been science included but also sociology and economics must be considered, and it is probably in these latter two areas that the major difficulties reside. Hence, no sophisticated aquaculture systems employing both' waste beat and nutrients have been developed at this point, but it has been probably only a matter of time before we will have extensive aquaculture using man's waste products.

Storage of Pollutants

Another method of controlling pullution has been is to store them at some distance from man's activities so that they have been as inaccessible as possible to the major fraction of the population and pathway to man. Occasionally nuclear waste products are being dumped in deep oceanic areas in the hope that when they do break out of their canisters, they will not find an easy parthway back to man because the deeper life forms have been probably not in any food chain involving man.

Perhaps not so well known has been the fact that deep ocean disposal is used for other materials such as average ordinance and poison gases for which the military has no further use.

In the case of nuclear products, there has been a very vociferous school of thought that believes storage facilities should allow waste retrieval if a technological break through is ever accomplished that would allow these nuclear waste products to be profitably utilized.

Asimov's suggestion has been to implant our hazarodus waste products in the oceanic trench areas so that within a few thousand years they will get carried into the interior of the earth and no longer be bothersome.

The only problem with this suggestion has been that the exposure time before the canister having the waste products is effectively buried in the crust is relatively long. During this time the canister seat has to be absolute. Whether this can be accomplished with present technology of presently being debated within the scientific community.

Nevertheless, storage has been still a viable method of controlling pollution, especially for the more toxic materials. At least theoretically, one knows where they are. The major concern with storage of toxic material has been accidental release, and the chances of this taking place always remain within the realm of possibility.

Controlling Pollution by Zoining

Another method of controlling pollution has been by specifying certain area that can be utilized for effluent discharge and other areas that cannot. Although we are even starting to convert our city dumps into usable areas, such as golf courses or ski slopes by covering, planting, and 'land-scaping as rapidly as possible, but it will be some time before ocean dumping areas are reclaimed in this fasion.

Certain coastal regions would be reserval for industry, others for sewage treatment, while still others would be set aside for recreation. Some of these activities have been such that they may get pursued in conjunction with one another, whereas others cannot.

The idea has been for every coastal area to try to get optimum utilization of marine resources available. In order to accomplish this type of zoining most effectively, a great deal more knowledge is needed than is presently available about the synergistic effects of various types of activities. One of the major advantages of optimum zoining has been that it it would allow maximum avoidance of undesirable synergistic effects.

The biggest stumbling block to coastal zoining probably has been inovercoming entrenched interests in regions where the optimum activities have been found to be different from those presently underway; a socio-economic problem rather than a technological one.

The concept is some what interesting one, because it essentially permits some pollution of certain areas while refusing to permit any pollution of other locals. This has been different from the way pollution has been presently controlled, where the effort has been made to limit pollution in all areas to the minimum possible amount.

With the joining concept pollution would still be limited as much as possible, but it might permissible to allow the water quality of ocean to degrade below the mean in some places, because in others it would be higher than the mean.

CONTROLLING POLLUTION BY TAXATION

One last method of pollution control is taxation which can be utilized to encourage or discourage certain activities in addition to raising funds to support government programms. Pollution would be allowed, but it would be taxed at a rate proportional to the environmental insult.

Some economists suggests that a tax on pollution will simply be a license to pollute and will therefore encourage pollution rather than discourage it. From a management point of view there have been a number of different directions that can be taken to control pollution, including both technological and socio-economic alternatives.

So far society has tended to aim more in the direction of the technological alternatives, but there appears to be a definite swing toward other methods. In the years to come it might be possible that more and more emphasis will be kept an these non-technical methods, especially as the tradeoffs, costs, and benefits become better understood.

FINAL REMARK

It has been said that pollution is a biological problem that can be solved by the application of physical and chemical methods. Whether this is a completely true statement is often a moot point, however, because the opportunity to apply any solution is usually won only after the employment of politics. Thus the study of pollution is a truly interdisciplinary one, including as it does the three Ps—pollutants, processes, and people.

IMPORTANT CONCLUSIONS

(*i*) Effective control of marine pollution requires further research work aimed at evaluating the extent of pollution of marine ecosystems, determination of the biological effect of toxicants and prediction of biological consequences. However, such an approach alone is inadequate, since the problem is urgent and suitable measures must be taken as soon as feasible.

(*ii*) Small marine organisms have a low resistance to toxicants. This fact was utilised to develop a system of rapid bioassays for immediate evaluation of the harmful effects produced by any given substance present in sea water.

(*iii*) The more promising applications of the results and methods toxicological research to the protection of the marine environment from pollution include: standardisation of the components of various toxicants in the sea based on biological criteria; ecotoxicological characterisation of sewage effluent and its components, for the determination of methods and norms of permissible dumping; recommendations on the priority and stepwise sequence of reducing the volume of sewage discharged by various industries; rapid tests of comparative toxicities of different substances, in order to select those which are ecologically safest; development and realisation of regional plants for protecting the marine environment from pollution.

Chapter 5

MARINE POLLUTANTS AND THEIR SOURCES

Once we have agreed on what pollution is, we can proceed with the specifics of the problem. Let us than in this chapter list and briefly discuss the more common pollutants appearing in the ocean, along with their sources.

As the list of pollutants is perused, it will immediately by seen that many of these are not necessarily uniquely produced by man but occur *normally* in nature.

However, in most cases, the activities of man have exaggerated the problem in the sense of causing a marked increase in the amount of pollutant present.

THE POLLUTANTS

In Table elsewhere in this chapter a list of common pollutants associated with the marine environment will be found.

These are not listed in any particular order of importance since it is found that in certain areas certain pollutants are more worrisome than others, while the opposite condition may be true in other areas.

This difference in importance results from local conditions such as population pressure, flushing, and climate. In later chapters the problem of why particular pollutants are more worrisome in some areas than in others will be addressed.

Table 5.1: Pollutants associated with the marine environment.

Pathogens	Toxic organics
Sediments	Petroleum
Solid wastes	Nutrients
Heat	Radioactive materials
Fresh water	Oxygen demand materials
Brine	Acids and bases
Toxic inorganics	Aesthetically displeasing material

Pathogens

The first entry in Table elsewhere in this chapter. is pathogenic materials. These are the living organisms that can produce sickness or biological unbalance in either plants or animals within the ocean itself or in humans who either contact oceanic waters or eat the organisms caught in the water.

These include a wide variety or *bacteria*, *protozoa*, *viruses* and *fungi*. The most common of these are normally associated with human waste products and are found in *sewage*. However, other pathogens may occur in non-waste disposal areas where environmental conditions are such that the proper conditions are present for *growth* and reproduction.

Sediments

Sediments are always present to some extent in the marine environment. Generally speaking, concentration of suspended solids increases with *proximity* to the coast, with the maximum occuring in the riverine and estuarine environment.

As man impinges on the coastal regions, his construction and agriculture efforts generally tend to increase erosion and therefore increase the amount of sediment in suspension. These sediments have a marked effect on plant growth since they block out a large portion of the light normally reaching *Sreator* depths and therefore decrease photosynthetic activity.

In some cases rooted plants are completely destroyed by this process. As sediments are deposited on the bottom they will cover up bottom dwelling organisms (benthos) such as oysters and may even smothes them under extreme conditions. In addition to destroying bottom dwelling life, deposited sediments tend to make the water shallower so that, in regions traveled by ships periodic dredging is required.

Dredging and the disposal of *dredged* material have their own set of environmental problems which will be discussed later.

The sediments of concern ate primarily *inorganic silts* and clays since the larger particles such as sand and gravel do not remain in suspenion for any monegligible peroid of time. They result from soil being washed from land areas into sea, in many cases because man has changed the character of the affected area by *agricultural* or *construction activities*,

Once the sediments settle to the bottom they may be resuspened in shallow water by natural causes, such as stroms, or by man as an operator of a power boat. The relative importance of recreational vessels in *maintaining turbid conditions* in *shallow water* is not well known at this time.

Solid Waste

The disposal of solid waste is a critical urban problem at the present time primarily because areas suitable for the dumping of the sevoluminous materials are becoming *scarcer*, especially when *transportation* costs are considered.

Due primarily to low transporation costs the ocean has been used as a dumping ground for solid waste for centuries. The solid waste most frequently marine dumped is sludge materials left over as a byproduct from domestic sewage treatment.

However the unused products of industry and the used-up products of society are also discharged into the sea.

If the product discharged at sea contains materials which may be leached into the oceanic environment, a serious problem could develop. especially if the materials are toxic. However, if the solid materials are inert, a ilttle *forethoiight* may result in some benefit to the marine environment.

Artificial fishing reefs, for example, have been very successful in certain area, especially for small fish. On the other hand, solid waste materials also can serve as a pollution in the sense of *destroying* habits when this materials is *dumped* into marsh or wetland area.

Until recent years marshes and wetlands were assumed to have no practical value, but recently we have found these areas to serve as breeding grounds for most of the commercial species of fish along with many other organisms important in the food chains of all marine creatures.

Thus there is a growing movements not only to save the wetlends from further destruction but also in many cases to attempt to rehabilitate there areas of extreme biological importance.

EXCESS HEAT

Excess heat, when added to the marine environment, changes ambient conditions, and these changes may be deterimental to the organism present.

The amount of heat that is *deterimental* and the extent of the degradation is determined by a number of factors to be discussed later.

The primary source of this heat is, of course, from electrical generating plants, whether they by *fossil* fueled or nuclear powered.

Fresh Water and Brine

Although fresh water may be in great demand ashore, too much of it in the ocean obviously will produce a marked environmental change within a small area. This change may be caused by properly designed storm *drainage* systems, water diversion *networks* associated with dams, and *effluent* of some *industrial* process.

Excessive fresh water usually is not a critical problem but it could be one if the effluents were directed into a sensitives area. The introduction of brine into the marine *environment* is similar in its effect to that of fresh water except in the opposite direction.

The organisms acclimated to a particular salinity now find themselves in a more saline environment which could cause peril anent damage. Brine is a *byproduct* of *desalination* plants, but changes in the *salinity regime* also can produced by changes in estuarine channel depths *necessated* by *shipping considerations*.

Toxics

Toxic inorganics are materials commonly used in industry, many of them *primarily* to control the growth of algae or to destroy pathogens. *Chlorine* is an excellent example of this In larger quantities this materialis relatively *harmless*; however, in larger quantities it may be quite *destructive*.

The key here apparently is how much and how often. There are petbaps 35 to 40 commonly used toxic inorgani cs and these should be very closely controlled by the user. The toxic organics are the most *disturbing* of the modern day chemicals commonly discharged either *purposefully* or accidentally into the *marine environment*.

These include the biocides such as fungicides, herbicides, insecticides, rodenticides, and also the additional organics including *halogenated hydrocarbons*, *petroleum* and *industrial chemicals*. The most disturbing toxic organics are the pesticides such as DDT and *ketone* which have the *unfortunate* characteristic of being more soluble in oil than in water so that they tend to *collect* within the fatty tissues of marine *organism*.

They are also very stable compounds which do not deteriorate very easily over a long period of time. These find way into the ocean both as *manufacturing effluent* and as runoff after *utilization*. Perhaps the most disturbing aspect of there toxic organics is that there are hundreds available with many more being produced by the chemical industry without any real knowledge of their long term effects.

Petroleum

Although petroleums may be classed as a toxic organic, it is a naturally occurring materials and is *biodegradeable*, given enough time. Petroleum in its natural llarge number very separate organic *materials* containing compounds, each of which seems to have it own effect on the biosphere.

These effects are not too well known and opinions within the scientific community range from one extreme to the other with respect to the toxicity of the material. It is a *aesthetically* disleasing of nothing else.

Petroleum finds its way into the marine environment due to accidents such as tanker damage or transfer loss, natural seepage, offshore production losses, losses associated with refineries, from runoff originating as *drippings* or disposal of used *automobile lubricants*, and unburned *hydrocarbons* emitted into the atmosphere as internal *combustion exhaust*.

Nutrients

Nutrients are what are commonly called fertilizers, those chemicals required by plants. The most common of these are the compounds of nitrogen and phosphorous. They exist in the natural environment at all times and some of them are recycled *naturally* by decaying diving organisms.

However, the activities of man have added to the total nutrient load of almost all cyastal areas. When nutrient levels get out of hand, plants grew *unchecked* so that decaying plants exist in such great numbers that the oxygen *supply* becomes rapidly depleted.

These nutrients are present in domestic sewage effluents, agricultural runoff, and the little understood but *apparently* important non-point source *runoff* from *urban areas*.

Radioactivity

Radioactive materials are not only discharged to the marine environ-ment by nuclear power plants, nuclear power plant fuel production and *reprocessing* plants, and *uranium* activities such as the burning of coal, when coal is burned it emits radioactive particles to the atmosphere which are then washed into the sea at a greater rate than any known nuclear power plants at this time.

Other sources of radioactivity are the natural backgro-und, weapons testing, mine *drainage*, *accidental spillage*, and a few isolated industries. One of the partially unsolved problems associated with the use of nuclear energy for electrical power is the long term storage or *disposal* of spent fuels.

This storage must be in an area such that there are no pathways back to man, and deep oceans have been suggested as meeting this criterion.

Oxygen Demand

Oxygen demand materials are those which require oxygen for degeneration and therefore steal oxygen which would normally be utilized by marine animals. Thus if too many oxygen demand materials ace placed in the marine environment, the animal population will be markedly decreased due to the lack of oxygen.

Sevage sludge and any other organic waste materials, even that resulting from excessive plant growth due to an over supply of nutrients, are examples, of common oxygen *demand materials.*

Acids and Bases

The discharge of acid and bases to the marine environment can be quite disturbing to the natural ecological balance of the systems. The normal pH of oceanic water is somewhere around 8.0, slightly basic. This is maintained by the carbonate system wherein carbon dioxide is moved back and forth through a *chemical* reaction to *bicarbonate* and *carbonin acid.*

If a large amount of acid or base is introduced into the system, the carbonate reaction will be offset and an important element of the environment will be affected. In addition, there are large *synergistic* effects associated with pH. Most toxic materials for example, increase their toxicity under condition of *low pH*.

The sources of *acidic* or *basic material* are *primarily industrial* with some of this material reaching the marine environment from accidental *discharges* and the *rupturing* of tankers.

Aesthetic Considerations

Aesthetically displeasing materials include all the stuff one finds in the ocean that is unpleasent to look at or to smell.

Tar balls, floatables, gas (often *hydrogen sulphide*) producing materials, and coloring, agents are some examples of pollutants affending the series, although in some cases these materials presents no real threat to the ecology of an area, when the area is being used for *recreation*, the quality of the *surroundings* becomes very important.

Thus even though an *aesthetically unpleasant* area may still be useable for many activities it is one which probably generates the most public *response* because on special instruments are required to measure the degradation; it is obvious.

We thereafter find that aesthetic *considerations* very often are more important to the average person than those that may be more damaging from an ecological point of view.

MANAGEMENT PROBLEMS

Even though the major pollutants have been declineated the management of marine pollution turns out to be quite complex. The reason for this is that the marine environment is used by man in many different ways this *multi-use* has a number of *ramifications*.

In the first place each oceanic activity pursued by man

may produce a whole spectrum of pollutants. A moving ship, for example, may not only discharge pollutants in the form of waste products from the people living aboard, oil from pumping its bilges, and heat from the discharge of its condensers, it may also require depending of a harbor channel.

This may produce a salinity change in addition to sedimentation problems associated with the maintenance dredging required. Each pollutant may come from many different users.

The nutrients which primarily seem to come from the effluent of sewage *disposal* plants also may come from strom drainage of urban areas in the form of waste products of domestic animals, along with runoff from agricultural areas that have been fertilized by either *commercial* fertilizers or farm animals.

Complicating the situation still further is the fact that pollutants sometimes affect each other. An interesting example of this occurs in baltimore. Maryland, a rather large, industrial city located on a small arm of the Chesapeake Bay.

A significant portion of the effluents. It turns out that the waste ferrous sulphate from the steel plant combines with the phosphate in the sewage plant effluent to from a precip tate Thus the phosphate ends up on bottom of Baltimore Harbor rather than dissolved in the waters of Chesapeake Bay.

In this case it appears that an *industrial*; effluent is utilized profitably (although accidentally) to decrease the input of a pollutant comprising one of the larger pollution problems in the *Chesapeake* Bay.

Another example of the effect of one pollutant on another is the attraction of heavy metals to suspended sediments. Most of the heavy metals in coastal waters seems to be swept out of the water by *sinking* sedimentary particles and end up in the bottom deposits.

This may or may not be desirable If the heavy metals

then find their way into bottom dwelling organisms which, in turn, find their way into a food chain pathway to man, then this obviously is not desirable.

A third example of this synergistic effect of two or more pollutants is the effect of oil on pesticides, combined with sediments.

As indicated previously, pesticides in the main are more *soluble* in oil then they are in water so that if there is oil present in *oceanic* water, it will tend to concentrate any pesticides that also may be present.

Oil also has an affinity for sediments and as it collects on the suspended particles, their weight is increased to the point where they sink to the bottom. Portions of the inshore marine environment thus are covered with sediments very rich in both oil and pesticides.

Any management activity therefore must be very careful in controlling one pollutant not to disturb the conditions of another so that a greater problem is not generaged as a result of some "solution".

The introduction of foreign material into the marine environment is not always detrimental to man or to the environment either.

The construction of fishing reefs was mentioned previously as one example of man's improving the environment by introducing foreign materials. Another example of this is the *generation* of *wetlands* in some instance due to man's *carelessness*.

In many areas the increased erosion produced by construction activities has resulted in large areas of *siltation* where these eroded sediments have bean deposited in the *shoreline* areas.

After stabilization, *populations* of marsh grass have been seen to spring up in many of these newly crated areas. In the main there has been no planning for this type of rehabilitation, but recent *experimental* work indicates that production of *marshlands* may very well be'a common tool of the *estuarine planner* of the future.

POLLUTION CONTROL COSTS

It should be constantly kept in mind that wastes are never destroyed. Waste materials, whether liquid, solid or gaseous, either can be moved from one location to a other on can have their form or chemical composition altered. But the material is still there.

Some authors have *suggested* the term "*waste placement*" be used instead of "waste disposal". Nothing is ever consumed ; it is just used for a while and then pushed aside.

Entropy is increased but there is no decrease in mass, and there is usually on increase in the amount of energy required to keep the system going.

This is why recycling is so attractive. The advantages in terms of energy and natural resource requirements along with the usual decrease in pollutants, are very strong. I This is why recycling is so attractive.

The advantages in terms of energy and natural resource requirements, along with the usual decrease in pollutants, are very strong.

The key appears to be cost, with must of the cost setting in the are of transportation. When the cost of recycled materials in low enough to be competitive with new material then recycling will be increased to the point where it becomes a a common part of every product that we use.

Cost is a major part of every pollution problem. For example, ocean dumping has been criticized by many to the point where the U.S. Environmental Protection agency is attempting to stop it completely.

In this context ocean dumping usually means dumping in the near coastal zone: within 10 or 15 miles of the area where these waste are generated. One satisfactory solution would be to require a distance of a thousand miles from the coast before dumping could proceed.

Here, the pathways back to man would be extermely

difficult and highly improbable. However, the cost of transporting this material such a great distance is, at this time, more than society is willing to pay so that this does not appear to be a viable solution.

POLLUTION SOURCES

Now that the pollutants have been delineated and described very briefly, let us concern outselves with the activities involved in the production of these pollutants in an attempt to pinpoint the source of the problem. These activities are listed in Table elsewhere in this chapter.

Marine Commerce

Marine commerce is deeply involved with ships, and ships require great many services. The first of these is a deep enough water environment so that they can pass to and for on their appointed rounds. This asually requires continous dredging, and in some cases plants are also to increase the depth of the presently available channels so that larger ships can be accomodated.

Dredging is an expensive proposition requiring large expenditures of money since the amount of material handled is large and the time in which this has to be done is relatively small in order to stay ahead of the sedimentation rate.

Once the sediments are extracted from the bottom and the channel is cut to the desired depth, the problem becomes one of disposing of the dredge spoils, the material dredged from the bottom. As indicated previously, sediments tend to have an affinity for a number of toxic materials, heavy metals, oil and pesticides being just a few.

Thus these spoils must be placed in an area where leaching back into the environment will be minimal. Various solutions have been suggested but the problem is still one that has not been adequately solved.

Another aspect of the ship problem is oil. Whether these ships are trasporting oil or using it for fuel and lubrication, there always seems to be some lost to the sea. Even though

Table 5.2: Pollution producing activities.

Marine commerce
Industry
Electrical power generation
Sewage treatment
Other non industrial wastes
Recreation
Construction

pumping of bilges is illegnl within the coastal areas, it may still be done of sea.

The oil input to the ocean resulting from bilge pumping is considered by many to be a major fraction of the petroleum products found in the ocean.

When the major purpose of the ship is to transport oil, the oil must be transferred from the ship to the shore holding facilities and *vice versa* and this process is also one contributing significantly to the amount of petroleum products lost to the sea.

Whenever ships are handled in a port area, a port facility is required. This facility must be maintained, improved, and modified to handle modern methods such as containerized shipping.

All this requires construction of one sort or another, and very often associated with this construction is some amount of pollution. Most harbors have large storage areas open to the atmosphere so that a certain amount of leaching into the marine environment is inevitable.

Lastley, there is a whole family of pollutants associated with any ship. A ship is just a little city afloat and any problems found in urban areas will be runoff on ships also.

These problems must be addressed, and solutions usually take the form of shipboard waste treatment or waste storage facilities capable of holding wastes util they can be transferred to shore facilities for treatment.

INDUSTRY

Whenever one mentions pollution to the average individually the first source of pollution that pops into mind is industry, and there is no doubt that industry does contribute some pollutant materials to the marine environment Just how much and how serious its contribution is a matter of some debate, but nevertheless most industries require the use of water.

This water has either heat or some chemical added to it before being discharged into the environment. These chemicals may be anything from toxic substances to harmless coloring agents. Electrical power generation bring forth vision of waste heat and certainly this is one of the byproducts of producing electrical power are other byproducts of elecrical power generation not so well know but just as important in terms of degreading the environment.

The cooling water utilized in electric power plants is sent through heat exchangers composed of many metal tubes. These tubes must be kept free of algal growth because if they are not, the amount of waste heat transferred to the cooling water will be drastically reduced, with an associated decrease in plant efficiency.

In order to keep these heat exchanger tubes clean, various methods are used. One of these is the occasional use of chlorine. If the chlorine is misused by employing large amounts in an attempt to overkill, this chlorine can very easily produce a large morality of organisms other than algae in the area of the power plant.

Present data indicate that more fish have been killed from excess amounts of chlorine that from heat discharged by electrical power plants. An additional problem involved with heat exchangers is the fact that many of these are made from copper, a toxic material used in antifouling paints on boats for many years, sometimes these heat exchanges are allowed to remain in an unused condition for extended periods of time with water sitting in the heat exchanger tubes.

Some copper leaches into this water from the tubes and when the plant is turned on again this copper rich water is pumped into the environment where it may cause some consternation among the local marine organisms.

Large amount of water pumped at high speeds from a coastal region can trap or entrain small or some times even large organisms. These creature may suffer mechanical damage as they go through the system, or they may be exposed to large amounts of heat for a short period of time.

Large individuals are particularly suceptible to the former conditions, while heat spikes seem to bother the younger ones the most. Thus mechanical and chemical as well as heat problems are associated with the running of the electrical power plant. All these problems must be considered together to produce a design solution resulting in an minimum effect on aquatic conditions.

Many modern plants have done this. However some of the older plants continue to insult the environment. Radioactive wastes have been cited as one possible disadvantage of nuclear power plants but the amount of radioactive wastes actually transferred by cooling water by the surrounding environment has been measured as extemely low and well within the limits set by the nuclear Regulatory Commission.

The spent fuel problems appears to be of more concern at this time. One industry presently relatively small but which promises to get much larger and spread itself out over much of the oceanic bottom is mineral extraction.

The pollution problems associated with this endeavour have not been documented due to the small opportunity for data taking, but it is expected they will be similar to marine constructions. At this time the major oceanic marine effect is directed toward the extraction of sand and gravel in the near by areas.

With in the imminent future it appears that a concerned effort will be made to extract the manganese noudles from the ocean floor, perhaps at depth as great as 3000 metres.

This mining effort will have some effect on the environment but how much is not quite clear.

Whether the bottom will be disturbed in areas important to the food chain of the fish eaten by man is still an unanswered question since, to a great extent, the bottom has not even been mapped for the major locations of those noudles.

If nothing else, however, one chain of effects will be to increase the number of vessels going to sea, requiring more port facilities to service them, utilizing more petroleum and producing more ship wastes. The same can be said for the increase in activity connected with offshore oil drilling although it appears that there, is less petroleum lost into the environment when the oil is obtained from an offshore well and transported to the continent by pipeline than when it is obtained from a land well and transported by means of tanker to its eventual user point. The spilage resulting from the use of tankers is greaters than that resulting from the use of pipe lines.

SEWAGE TREATMENT SYSTEMS

The treatment of domestic sewage appears at first glance to be an easy ii ; this is a process that has evolved over a period of many years and certainly by now one would expect the process to be reasonably well developed.

The process treating human waste products is quite successful in removing suspended materials and pathogens, but it is extremely difficult to extract all of the nitrogen and phosphorous compounds. Consequently sewage treatment facility effluent contains resonably high amounts of these nutrients which push plants toward uncontrolled growth.

In addition, when the facility is overloaded, as during a heavy rain, it is often completely bypassed ; discharging raw sewage directly into the water. This requires a great deal of oxygen for decomposition so that the biological oxygen demand is increased. The largest problem facing

Table 5.4: Estimated annual input of oil to the oceans from maritime transportation activities.

Source	*Million tonnes*
Load-on-top tankers	0.31
Non-load-on-top tankers	0.77
Bilges and bunkering	0.50
Terminal operations	0.003
Dry docking	0.25
Tanker accidents	0.20
Non-tanker accidents	0.10
Total	**2.133**

sewage disposal plants at this time is probably that generated by the fact that many of the sewage system customers are not homes but industries. Some of the industrial effluents flowing into sewage treatment plants are difficult, if not impossible, to treat adequately.

It is primarily for this reason that sewage sludge is so difficult to dispose of, since it very often contains large amount of heavy metals and other toxic industrial wastes. One aspect of sewage treatment neglected up to now, primarily because the cost are so over whelming, is the treatment of storm runoff.

Unfortunately the solution is not as straight forward as simply increasing the capacity of existing plants to handle the heavy loads associated with storms, or even to odd treatment to storm drainage systems when they exist separately.

Much of the urban and rural runoff does not go through conventional pipe systems, so additional systems would have to be built it all run of is to be treated. Storm drainage washes urban streets clean. All material deposited on the streets, including crankcase drippings, rubber worm from auto trier, and bird, dog, and cat droppings are washed to sea.

Modern studies have inlicated that this runoff is a very

significant source of coliforms and petroleum, to name just two pollutants of some concern to man. How to handle these nonpoint sources is a task presently facing marine environment managers and one that appears to have **no** inexpensive or simple resolution.

With most large urban areas disposal of the solid residue from sewage treatment plants is a solid waste disposal problem of the first magnitude.

Ocean dumping in the near-shore coastal regime has been utilized for many years and, to a certain extent, will probably continue to be. Either the sludge must be treated more thoroughly before it is dumped or new deep sea dumping areas must be found. It new treatment or disposal techniques are not followed, there may be major changes in the coastal environment.

AGRICULTURAL SOURCES

The largest single class of nonindustrial polluters are farms. Agriculturat pollutants include commercial fertilizers, animal wastes, pesticides, herbicides, and sediments. In many cases the agricultural contribution of all these pollutants in greater than any other single source.

The pollution can be stopped, and there have been cases in the past where particularly bad problems have been solved. An excellent example of this is the case of the duck farms on long island.

Long Island duck is a delicacy that Americans have enjoyed for many years so that the Island has developed a very large duck industry.

For years the waste products from these ducks had simply washed into Long Island Sound and the incremental effect on Long Island Sound was getting more alarming each year.

The sound was becoming completely overgrown with undesirable plants from the excessive nutrients, and all of the local fisheries were suffering very badly.

A concerted effort was made to clean up the duck farms by venting direct runoff from the farms and collecting the duck faces. This effort was extremely successful and Long Island sound fisheries are being steadily rejuvented.

Recreational Activities

Usually when we think of marine recreation we think of a nonpolluted area since recreation is essential limited by pollution. We very rarely think of recreation in terms of producing pollution, but there is a significant amount of pollution produced by recreational activities.

In shallow water motor boats stir up the sediment on the bottom, a conditton which often persists for long periods of time. This increase in turbidity markedly affects the natural plant growth and probably has something last few years.

In addition, power boats produce was, especially when operating at higher speeds, which tend to accelerate the rate of shore line erosion. This is especially true in narrow rivers and estuaries. The aspect of motor boating that has many investigators most concerned is the discharge of petroleum products into the environment.

There are some figures indicating that recreational boats add as much pertoleum and petroleum waste products to the environment as all other sources combined. Even fishing sometimes adds to the pollution problem. Many fisherman, if they catch a fish of an undesirable type, will simply throw it back.

The fish, as often as not, has been mortally wounded and will die. If the area is heavily fished, these dead organisms will have a significant impact on the environment as they got. As an aside, it should be noted that over-fishing is just as effective in decimating a fish population as any toxic pollutant.

CONSTRUCTION

The last source of pollutants listed in Table elsewhere in this chapter is construction. By then done on hand contiguous with the sea, such as port facilities, Implied here is both the actual construction of the structure and the associated activities that go along with construction.

There must be all sort of support equipment, but most important, there must be all sort of support equipment, but most important, there must be support people Since, in general, the amount of pollution is directly related to numbers of people, any large construction effort will bring with in the usual people waste products.

The effect of a new physical structure were one had not previously been is not too well known, but one thing is sure the environment will never be the same as it was before. Some of these changes may be beneficial, but no matter how beneficial there are always going to be some members of society who real to any change by claimbing a degradation of the environment.

Any marine construction project must be judged on the basis of the benefits to be derived by society as a whole rather than *solely* to those accruing to some small group.

Sources and Quantities of Marine Oil Pollution

The first step in assessing the threat arising from ship and off shore production is to see it in perspective against the other the *apparently* less emotive sources of oil. In 1978 Cowell updated earlier US National Research Council data Table sources of oil in the *marine environment. Comparison* with the earlier 1973 Us data.

Shows that considerable improvements were made in reducing ship source pollution in the '*nterventin period.* The Intergovernmental Maritime Organisation, IMO (formerly IMCO) is currently engaged in another review which is expected to show further improvements in the period up to 1982.

However, in terms of the perspective which is to be established it must be concluded that land-based sources contribute more to marine pollution than do ships almost three-times as much that the contribution from off shore oil

Table 5.3: Estimated annual inputs of oil to the oceans.

Source	*Million tonnes*
Land on-top tankers	0.11
Non-load-on-top tankers	0.50
Bilges and bunkering	0.12
Terminat operations	0.001
Dry docking	0.25
Tankery accidents	0.30
Non-tanker accidents	0.10
Off shore oil production	0.06
Coastal oil refineries	0.06
Industrial waste	0.15
Municipal waste	0.30
Urban runoff	0.40
River runoff	0.40
Natural seeps	0.60
Atmospheric rainout	0.60
Total	**4.95**

production is comparatively very small, that accidents contribute about 15% to total marine source pollution and only about 5%, of the total oil input to the sea. Such surveys provide very useful information on the relative amounts from various sources, some of which otherwise receive little or no publicity, and helps in this way to provide a better perspective.

As we have seen, they also provide a useful record of improvements when this has occurred. On the other hand, global figures in themselves do not give a direct estimate of possible effects. Thus if we take atmospheric rainout as possibly the best example of a uniform input, its impact is likely to be less noticeable than localised discharge through rivers.

Again, the contribution of *hydrocarbons* from a river may be negligible as a contribution to global figures but quite important in may be expected to have a greater effect of hydrocarbon burden in a river may be expected to have a greater effect in the river itself then in the sea to which it discharges.

Of course, as stated earlier, it will readily be assumed that massive releases of oil will have a much greater effect of hydrocarbon burden in a river may be expected to have a greater effect in the river itself than in the sea to which it discharges.

Of course, as stated *earlier*, it will readily be assumed that massive releases of oil will have a much greater effect than low. concentration operational discharges. Thus global figures are all very well, but in order to assess actual impact in a given location, one needs to consider on the one hand the presence or absence of continues oil layers on the sea surface, and on the other, the *concentrations* of *hydrocarbons* in the water column itself.

Chapter 6

THE EFFECT OF MARINE POLLUTION

Clearly the opportunities to study the effects of oil presented by continuous natural seepage, refinery effluent, oil rig discharges, etc. on the one hand, and the more massive short term injections of oil arising from oil rig blowouts and tanker accidents on the other, are different, and differences exist even within these two broad classes.

It may threfore by through that results obtained in one case may not be applicable in the others. It is thus worthwhile to consider types differences and to assess the extenC t to which results obtained as the various sites and incidents may be of universal application. The first point to note in this endeavour is that oil does not dissolve in water.

It floats on the sea surface as a thin layer when spilled or otherwise discharged because it is less dense than water and insoluble in it. This thin layer tends to break into droplets by wave action and these droplets become dispersed in the volume of sea water beneath the slick.

Photochemical and biological oxidation of the sicks and of the dispersed oil droplets takes place producing oxygenated derivatives of the original oil components and these may form true solutions in water. Thus the area of sea covered by the surface slick is proportional to the magnitude of the spillage but the concentrations in the sea arise from the slow dispersions of droplets of oxygented derivatives.

Sea water concentrations are therefore a function of mass transfer rate from slick to water and not a function of the quantity spilled. Sea water concentrations are in fact independent of the quantity spilled, provided a surface layer has formed above the sea water volume in question.

If more or less oil is spilled or otherwise introduced to the sea the result is more or lies sea affected rather than that a given volume is affected to a greater or lesser extent.

Oil concentrations are therefore controlled by the in solubility of the oil and mass transfer rates from sarface slicks to the underlying sea.

In the absence of detailed knowledge of mass transfer rates it is not possible to calculate the resulting concentrations, but upper lim.ts can easily be established.

Thus if the slick thickness to which the oil spreads and from which it subsequently disperses to the water is 0.1 mm, then the maximum concentration which can achieved if the oil were instantaneously dispersed uniformly in the top meter of sea would be 100 ppm. Concentrations over other depth ranges and for other thickness are of course pro data.

It must be calculated therefore that is we leave aside for the time being the possible environmental effects of continuous oil slicks on the sea surface which occur when the capacity of the acqueous phase for oil is exceeded, we can discuss the other oil concentration-related aspects in the same manner whether a surface slick is present or not and whether the oil in question across from a natural seep, blowout, cargo tank repture or operational discharge from ship, a oil rig or refinery.

Generally speaking, we have limited our discussions to those materials introduced into the marine environment that ale the direct result of man's activities, and with a few minor exceptions we shall continue to limit our discussions of pollution to those activities of man and not consider the deletrious effects of natural occurences.

POLLUTANT TYPES

In this chapter we shall attempt to get some idea as to what pollution does and why we consider it to be so undesirable. There are three basic type of pollutants : the pathogenic, the aesthetic, and the ecomorphic.

Pathogenic pollutants are those which cause disease. This disease may be fatal if the pollutant is a lethal poison. Aesthetic pollutants are those pollutants causing a change in the environment displeasing to the eye, ear, or nose of man.

Ecomorphic pollutants, on the other hand, are those pollutants which produce a change in the physical characteristics of the environment in such a way that there may be drastic changes in the structure or composition of the biosphere. Obviously, these three types of pollutants are not of equal import.

Oathogenic pollutants are certainly much more serious than the other two; however, ecomorphic pollutants are often an indicator of more serious types of pollution to follow.

Further. more, since pollution is man-produced and the effects are suffered by man, aesthetic problems are certainly of interest.

Nevertheless ih any management situation when trade-offs are required, it will certainly be necessary to rank different types of pollutants so that society can determine how it would like to spend is limited funds of time, effort, and money.

Acuate Effects

Let us look first at the pathogenic pollutants since these are obviously the most important. The effects of pathogenic pollutants are either acute or chronic.

Acute effects are those which occur in a relatively short period of time while chronic effects may take time to be noticed, as in the case of carcinogenic pollutants, or the effect may be noticed immediately but continued for a very long period of time. The accute effects are the easiest to determine because of the short time between administration and affiction.

The easiest way of measuring acute effect if by feeding the pollutant in question to test organisms, such as since or selected, fish, and observing the dosage required to will 50% of the organisms involved. This is usually done by feeding a pollution of test organisms increasing amounts of the pollutant, and for each dosage the number of individuals that succumb is noted.

When these data are plotted, an S-shaped curve results, such as shown in Graph -3.1 wherein dosage versus number of mortalities is plotted. From this curve it may be seen that there "re a small number of individuals that require an exteremely large does before they are killed while some are killed with a relatively small dose.

By plotting this type of curve, it is leasy to locate the dose for which half the individuals die. This dosage is called the LD 50, and it is this LD 50 which is usually used to compare the relative toxicity of one pollutant with another. Another way of examining acute effect is to observe the effect of a given pollutant concentration on organisms over a somewhat longer period of time, noting how long it takes for given dosage to produce mortalities.

This say be done by giving a predetermined concentration of the pollutant to the test animals and observing how long it takes to kill half of them. When data from an experiment such as this are plotted, a curve similar to that shown in graph 3.2 results.

As may be seen, the curve is a hyperbola with the horizontal asymptatic suggesting a threshold concentration. In other words, below they asymtotic value there is no apparent effect, since at this concentration an finite amount of time is required for mortality. The vertical asymptote, on the other hand, appears to be a saturation level.

At this level any increase in the concentration will not produce any further decrease in the amount of time required for fatality. Evidently there is a finite time required for the pathogen to do its work, so that no matter how high the concentration, the time required to kill will not go below this amount.

Effects

With the asympototic values, suggests another concept ; the threshold velue. There appears to be a dose size for each pathogen below which the pollutant has no effect. The concern with threshold values of various pollutants is one aspect of the study of chronic effects Chronic effects are the most difficult to measure because of the time involved; in some cases cancers have developed as long as 25 years after initial exposure to the carcinogenic agent.

Nevertheless, measurements have been made and for many materials threshold value seems to describe the data reasonably well. Most experiments need some measure of the effect on the organisms other than death, so that many different measures are used. These are; deformity in the growth of organisms, damage to particular organs, change in heartbeat or breathing rate, genetic damage to individuals, change in the rate of increase of a population, life expectancy of individuals within a particular population, and fecundity of females with the population.

Unfortunately, and index may be very descriptive of the effect of a pollutant on a particular species, but it may be completely ineffective in describing what happens with this pollutant comes in contact with other species. A typical experiment using one of these measures would involve the variation in pathogen concentration while noting the variation in the hearbeat or some other measure.

When these, data are plotted the result is very often a straight line as shown in graph 3.3. This linear relationship between effect, and concentration may be extrapolated back to determine the concentration for zero effect. When this is done the line will cross the axis at a concentration from zero. This is called the threshold value.

A great deal of caution must be used in interpreting

these data. Some pathogens apparently have no threshold value associated with them, and it is always risky business to extrapolate a straight line. The actual manner in which the organisms reacts to the stimulus might veiy well be non-linear at lower dosages.

In addition some chronic effects may be so saitable that two or three generations of the organisms might be required to observe any change.

Synergism

Another complication in the examination and quantification of pathogenic pollutants is the phenomenon of synergism. A synergistic effect occurs when the combined effect of two or more materials acting together is greater than would be expected from the simple sum of the individual effects.

Temperature, for example, is a very strong synergistic parameter, generally enhancing most pollutant effects as temperature is increased. The same is true of salinity and dissolved oxygen. Some pollutants, on the other hand, tend to decrease the effects of others, as might be in the case of a strong acid and a strong base present at the same time. Other pollutants may cause strong pathogens to be precipitated out of the water column or to be collected on sediments, so that in either case they end up on the bottom.

Thus it is extremely important in the analysis of any pollutant to be aware of the other materials present and their synergistic tendencies. Within the marine biosphere many organisms are affected by various pollutants. Whether or not a particular organism is affected by ony given pollutants will depend on any variable.

For example, many polluants are found only in the water column while others, such as some of the pesticides and heavy metals, are adsorbed onto suspended sediments and usually find their way to the bottom within a relatively short period of time.

Consequently, these latter types would tend to affect those organisms living on the bottom or feeding from bottom organisms more then they would affect the organisms living within the water column, such as free swimming fish. Also of import in determining the effectiveness of a particular pollutants on living organisms is its relative solubility in water and oil.

Pollutants more soluble in oil will tend to find their way to organisms that have a larger content of oil in their body tissues. This is particularly true of some of the pesticides which are more soluble in oil than water. The individual habits of marine organisms also affect their susceptibility to a particular pollutant.

Feeding habits with respect to time of day, portion of water column for aged, type of material ingested, and the method of digestion utilized all determine to a great extent the types of pollutants the organism will be exposed to. Similarly spawning habits determine the type of area and time of year in which a species is particularly sensitive to subtle changes in the environment.

These changes affect not only the parents but also the offspring, and may be even their progency. Other habits not having to do with spawning also help to determine whether an organism will be exposed to particular pollutants. Wheather the organism is sensitive to light or sound, for example, will often determine whether it is attracted to a particular out-fall.

Thus we find that a pure and simple cause and effect relationship between a pollutant and an organism is not enough to determine completely the actual response of the organism in the real world.

Laboratory results might be completely refused in the field simply because the organism does not respond to extraneous characteristics of the environment under natural conditions in the same manner in which it did in the laboratory.

Pathways to Man

Once some of the pollutant has been ingested by marine organisms, the major concern of man is the possibility of this pollutant appearing on his dinner table and causing a pathogenic response. Obviously if a fish ingests a pollutant and a human eats that fish, then that person will have ingested a dose of that pollutant and there will probably be some danger. The question is, "How much ?"

In order to arrive at an approximation of the actual amount of undesired material finding its wax into the reader's stomach one must follow the pollutant through a number of steps. Most organisms living in the sea tend to concentrate materials existing in the sea to much greater values that they are in the ocean itself.

For example, a diatom, a small form of marine algae having a slicate frustule, has a body concentration of silicon about 40 thousand times as great as the oceanic waters from which the organism derives all of its materials.

Thus the diatom concentrates silicon very effectively so that if we eat diatoms, a large portion of our diet would be grassy. Similarly, other organisms, including fish, will concenrate other materials even if these do not naturally occur in the environment.

But this amplification or concentration does not stop with one step. We must continually keep in mind the fact that an org inism does not exist by itself in the marine environment.

Any organism is dependent upon many other organisms for its existance. An edible fish, for example, might very well subsist on smaller fish which, in turn, might subsist on small zooplankton, which, in turn, may subsist on phyto. plankton.

There may be 5 to 10 individual links in a food chain leading to man, and there is the distinct possibility that for each one of these links a concentration of the pollutant occurs.

Thus the total pathway to man for the pollutant might be a rather tortuous one, but it might result in a very high codcentration of the pollutant material in the fish.

This in itself may be of no concern to the individual unless he ingests the fish and if so, how often and how much he consumes. It is not enough simply to know that there is some pollutant present in marine organisms: one also should know how much is present, and in addition two other facts are required.

This first is the accepted maximum level of ingestion of the pollutant before harm results, and the second is the amount of fish that can be eaten before this level is reached. In this way it might very well be possible to have a particular kind of fish one or two times a year with perfect safety, while if it is eaten once a day, it might very well have a toxic effect.

Ecomorphic Pollutants

There are many different types of pollutants that may be classified as ecomorphic while some pollutants fall into all three of the classes stipulated above and many fall into at least two. Sediment is a pollutant whose effect is primarily that of changing the environment since it is not pathogenic and very often is not aesthetically displeasing, although it can be.

As it settles to the bottom, it will tend to bury the organisms that live on the bottom, such as cysters and clams, and eventually smother them Sediment also tends to fill in marsh areas, killing the marsh grass, and completely changing the habitat.

Since marsh areas are breeding grounds for small fish and an important ecological link in most of the oceanic life, any destructive process such es this is bound to markedly change the marine biological environment. On the other hand, in some cases where man's activities have produced large amounts of erosion and sediment has been brought down by rivers into estuaries, some estuarine areas have been filled enough to support marsh grasses. In some cases excess erosion can produce marsh lands rather than destroy them.

Whenever there is a depth change in an estuarine or

coastal area, the effects are many. They range from changing the biological environment so that larger fish will no longer live in the shol areas to affecting the total circulation pattern of the area.

But *suffice* it to say that changing depth in many cases does change the manner in which salt and fresh water mix in the coastal area and also the amount of flushing present in a particular region. When the sediment is in suspension it also has an effect on the eco-system, the most obvious aspect being a change in transparency.

By making the water more opaque, light will not penetrate as deeply as before so that the thickness of the layer of water in which plants grow will be decreased. If the water is fairly shallow to begin with, this decrease in transparency may very well result in the inhability of rooted plants to grow and some marsh lands may be destroyed.

The effect of sediment, then is two-fold: changing both bottom and water properties, Both of these effects tend to change the physical characteristics of the ecosystem. Another ecomorphic pollutant is oil, although oil sometimes products toxic effects.

The result of oil coating bird feathers is a well known physical effect. The oil does not poison birds ; it4 $ 'makes it impossible for them to fly and consequently many of them die. A film of oil on the surface will also decrease the amount of sunlight entering the water and, similarly, will limit the exchange of oxygen from the atmosphere. In addition to the aesthetic and pathogenic effect pf oil, there are thus also ecomporphic effects.

The toxic effects of various nitrogen compounds are well known but fertilizers and other nutrients washed into the sea hove a non-toxic effect on marine waters. These materials will do just what they are designed to do: make plants grow, whether these plants be terrestial or aquatic plants.

The toxic effects of various nitrogen compounds are well known but fertilizers and other nutrients washed into the sea have a non-toxic effect on marine waters. These materials will do just what they are designed to dot : make plants grow, whether these plants be terrestial or aquatic plants.

Those plants on the bottom of the blanket will tie, due to lack of light, and those roting plants will utilize oxygen in their decay processes. Even though there are more plants than before, producing more oxygen in the process of photosynthesis, so many of them are dying and decaying the result is to decrease the dissolved oxygen.

This phenomenon it called "Eutrophication." Another ecomorphic pollutant that seems to be receiving a lot of attention lately is heat. Usually supplied to the marine environment form electrical power plants or other industrial sources, this heat will increase the water temperature and may result in different kinds of damage to marine organisms.

Since most mutations are caused to thermal activity under normal conditions, the possibility of genetic damage is real. It is estimated that about 90% of the mutations occuring normally are thermally caused, while only about 10% are caused by radiation.

Aesthetic Pollutants

The last type of pollutant mentioned above was aesthetic pollutants. Aesthetic pollutants are those which offend the human senses and as such are considered by many to be unimportant. In many environment problem areas, though, the aesthetic considerations often receive more attention by environmentally oriented citizens than some of the pathogenic pollutants.

This is simply because the aesthetic pollutant are visible and what can be seen, smelled, or heard appears much more bothersome than that which is hidden, even though the hidden pollutant might be much more dangerous to life. Consequently, any scheme for controlling marine pollution must include some consideration of the aesthetic pollutants if it is to succeed.

An excellent strategy seems to be to focus on the aesthetic pollutants and use them as a machanism to gather momentum to clean up the rest of the pollution problems.

Power transmission cable has caused more than one power company grief because of the environmental consideration given the local citizens who are concerned about the appearance of power cables running across the countryside.

Oil wells, especially the offshore variety visible from resort areas, are of great concern to most environmentalists. The same thing is true of equipment required for dredging operations and the spoil areas resulting from these operations.

To some people, especially those living in relatively pristine environments near th, water, boats can also be an aesthetic pollutant, especially those that cause waves and noise.

Effluents that cause the colour of the sea to change to increase the amount of suspended sediments also may be considered aesthetic pollutants. It simply is not as pleasant to go swimming in water that is cloudy of contains colouring material different from what one would expect in a normal, natural body of water.

In essence, any change from the pristine environment appears to many to be aesthetic pollution Obviously we are living in an unreal world if we think the marine environment can be maintained in an unspoiled form.

It sbould be borne in mind that many complaints about aesthetic pollution come from individuals who would father live in the primeval environment not realizing that it would not be available to them unless the commercial world about them supported it.

There appears to be a dichotomy wherein of try to maintain the unspoiled environment in certain areas while letting others become over utilized Nevertheless, there is a growing movement to include in the cost of every project a relatively small sum to improve its aesthetic qualities.

This is, of course, exactly the same philosophy that is applied to the control of other froms of pollution. Additional money is spent in a chemical process, for example, to control the pollutionproducing propel ties of the effluent since society believes that this additional sum of money is desirable expenditure.

It may similarly be maintained that the expenditure for aesthetic pollution control is desirable as it also will tend to increase the quality of life. This paragraph briefly those portions of these studies that are relevant to determining the effects of oil on marine wetlands.

Also, the literature and out personal experiences with oiled marine wetlands are synthesized to allow an evaluation of methods or protection, cleanup, and restoration attempts that have been carried out in matine wetlands. This section accomplishes there two objectives by:

— Providing a brief review of the effects of all spills and related cleanup activities on salt marshes and mangrove ecosystems.

— Reviewing methods of protecting marine wetlands , from being oiled.

— Reviewing successful means of cleaning marine wetlands following oil spills.

— Reviewing and presenting techniques that have proven successful in restoring marine wetlands damaged by oil spills and/or cleanup operations.

— Establishing a set of criteria and discussing guidelines for decisions on means of protecting susceptible areas, and for cleaning and restoring oiled marine wetlands.

— The differences in growth patterns and life cycles are essential elements in consideration of methods of cleanup and restoration of marine wetlands.

— It should be noted that salt marsh and mangrave vegetation occur together in some parts of the world, *e g.*, souther Brazil, souther Australia. For the purpose of this discussion, we need to identify two terms.

— Recovery : return of a site to dominance by native organisms that are within the natural range of limits for structure and function in unprotected example of the ecosystem within the local geographic area.

— estoration : man's efforts to initiate and/or enhance the recovery process.

EFFECTS OF OIL SPILLS ON MARSHES AND MANGROVES

Oil Impacts to Marshes

Oil impacts to marshes will very with a number of factors including the amount of oil, type and magnitude of cleanup activities, type of oil, physical and biological structure of the marsh, laitude, and season.

Subsequent persistence of damages and recovery of the system also depend on numerous physical, chemical, and biotic factors including weathering rate and degree of removal or retention of oil, availability of propagules (seeds or vegetative), successional processes, sediment erosion/ acceration, and restrorative activities by man.

Plants can be affected in several ways. Physical smothering by oil can lead to reduced transpiration, respiration, and photosynthesis. Absorption of toxic oil fractions through the leaves or roots may cause poisoning of the plant by disrupting cell membrances and cellular organelles.

Some species such as *juncusmaritimus* is *gereraii,* and Spartina anglica exhibit epidermal or cellular resistance to oils. There is a considerable range of sensitivity to oil. Some species are resistant even to repeated spills (e.g., salicornea sp. Many of the more resistant species are also pernnials with large root systems that enable them to grow back following oil damage to aerial portions.

Oils vary in their toxicity, according to the content of lowboiling compounds, unsaturated compounds, and aromatics. In general, the higher the concentration of these constituents, the more toxic the oil.

It is self evident that the amout of oil and frequency of

oiling will effect the degree of damage, but it is less obvious that visocity of the oil is important. With light oil, there is a physical limit to the amount that will "stick" to a shore, with subsequent tides redistributing oils.

In contrast, the viscosity of heavy oils often allows retention of large volumes and many cause mortality not through toxicity but through smothering. Heavy oils also tend to be more persistent and to extend the durtation of damage, thus slowing recovery.

Since and sediment structure of the marsh considerably affect the degree of damage and potentials for recovery. In the United States, Europe, and elsewhere marshes may be narrow fringing bands along the high water mark, especially in areas where tides are small, or marshes may

Table 6.1: Generalized responess of mangrove forests to 00 Spills.

Stage	*Observed impact*
0-15 days	Death of birds, turles, fish, and invertebrates.
15-30 days	Defoliation and death of small [< 1 m] mangroves-loss of aerial roots.
30 days-1 year	Defoliation and death of medium [<3 m] mangroves-issue damage to aerial roots.
1-5 years	Death of large [< 3 m] mangroves, loss of old aerial roots and regrowth of new ones (sometimes deformed). Recolonization of oil damages areas by new seedlings.
1-10 years	Reduction in litter fall, reduced reproduction, and reduced survial of seedings.
	Death or reduced growth of young trees colonizing spil site?
	Increased insect damage?
10-50 years	Complete recovery.

be broad vegetational belts (occasionally many hundreds of meters wide) in estuaries or on sheltered coasts, especially where tidal ranges are large.

Extensive mashes may not be completely covered even to massive spills, as oil will tend so affect the sides of the creeks and drainage chanaels first and then progress into the marshes plateau. Where vegetation is tall and dense (e.g., in a spartinetum of juncetum), considerable volumes of all may be completely oiled, and after severe damage, there may be no source of propagules for recovery.

In such areas, restorative action, may be the only means of marsh reestablishment in the short term. Sediment characteristics, including particle size distribution and drainage patterns, considerably affect potential for oil retention.

Marshes in North America and Europe can be found in sediments ranging from medium grainted sands to very fine silts where more than 95% of sediment particles are less than 63 in diameter. Sands tend to be more freely drainging and more easily penetrated by oils as shoots can occur.

If oil penetrates anaerobic sediments„ weathering is reduced or stops, and the oil may be retained in unchanged form for considerable periods of time, reducing or preventing plant regrowth. However, where oil is weathered before or after incorporation, even very high levels may not prevent plant regrowth.

Growth in salicornia sp., *suaeda marivima, Aster tripolium and Spartina anglica* has been reported run areas where levels of total aliphatic hydrocarbons of up to 5,875. μg/g were found. Seasonal changes in the marsh may considerably effect the extent of oil damage. Oiling before or duridg flowering can cause reduced flowering and seed production.

This may delay flowering and seed production. During winter months may species die back and oil may affect only dead and decaying vegetation ; however, in spring and summer growth, new shoots and germinating seedflings are particularly vulnerable. In some cases, a major problem following oil damage to vegetation has been loss of sediment stability followed by erosion (e.g., Isle Grande, France following the Amoco Cadiz oil spill).

Where extensive rhizomatous root mats ramify the sediments or where vegetation cover is continuous (e.g., in denise Spartida Stands), sediments may remain stable even when vegetation has been killed.

In low marsh, colonized mainly by annual species (e.g., Salicornica) or where vegetation is sparse, root mats may be insufficient to prevent erosion. Where erosion occurs, lowering of marsh level may prevent reestablishment of vegetation Erosion may also follow damage to creek systems after cleanup activities (e.g., during the Amoco Cadiz spill that affected the Isle Grande marshes.

The importance of slat marshes in prevention of coastline erosion is well established and may be or prime importance in some areas. Because of the preceeding considerations, basin forest may have variable characteristics. Generally, however, basin forests are well developed with canopy heights that may reach 15-20 m.

Riverine Forests

This forest type develops along the edges of estuaries, often as far as the inland extent of the saline intru. sion. It is best developed, however, in the lower course of the estuary where water flows are moderate to high and high nutrient inputs prevail. This combination of adequate freshwater and nutrient inputs allows the development luxuriant stands.

On the periphery of the forest, the dominant species is the Rhizophora mangrove. This is an area where there is greater kinetic energy by due to the combination of river flow and tidal motion. The complex root system of Rhizophora, with its extraordinarily developed adventitious ("stilt") roots, allows the establishment of large trees on *very* unstable, soft soils.

In the inner forest, mixed stands of Laguncularia and Avicennia occur. Rieverine forests are highly productive with litter fall rates in excess of 4 g/m/day. Canopy heights reach 20 m in the best-developed forests.

Fringe Forests

This type occurs along the seaward edge of forests lining protected shorelines. There is usually a strong horizontal gradient in tropography as well as a singificant gradient in turbulence and tidal amplitude.

Every derived from the tides and waves decays rapidly inland where tidal flooding is further reduced by the rise in the mangrove floor. The reduced flusing inland results in the establishment of a physiochemical gradient in water quality, toward the other edge of the fringe, where greater energy levels prevail, the large, heavy seedings of Rhizophora become established. Inland this way and tidal energy dissipates, and in the higher, less flushed parts of the fringe, Avicennia becomes the dominant species.

Since fringe are exposed, they are subjected to periodic destroction buy storms, waves, or scouring by strong current as. During storms large amounts of debris may be deposited in the outer fringe, reducing circulation to the inner fringe and the basins inland.

Overwash Forests

These forests are formed on small island or projections from land masses (peninsulas) they are characterized by intense flushing by daily tides. The high flushing rates do not allow the formation of strong salinity gradients nor the accumulation and aggradation of the soil or of nutrients factors that usually lead to a succession of species toward the interior.

True overwash forests are thus monospecies (cominated by rhizophora). The amount of flushing and nutrient availability are critical factors that determine the structure and the rate of overwash forests.

They, overwash islands in coastal oligotrophic waters may not be as developed as those inside coastal lagoons where higher nutrient accumulation is possible, excessive flushing leads to erosion and loss of the forest whereas extremely low flushing rates lead to accumulation of debris and vertical aggradation inside the forest with reduced flushing rates toward the core of the island. These islands thus evolve into fringe and basin forests. Cintron et al. have described the formation, aging, and decay of these islands is southwest Puerto Rico.

Dwarf forests, Dwarf forest are found in marginal environments where their structural development is limited by edaphic factors, probably the unavailability of nutrients.

Mature trees in these forests are usually less than 3 m tall. Salinity is not the limiting factor. Although the vegetation is limited in heigh, leaf size is not reduced as usually happens when mangroves are under the influence of a chronic stresson. This type of forest is different from those that may be considered scrub forests and that develop under the influence of chronic stressor, such as high salinity or cold stress.

The dominant species in Dwarf forest in Florida and Puerto Rico is Rhizophora. The Table elsewhere in this chapter List 14 oil spills, worldwide, where some documentation of impact to mangroves has been observed and reported. Most of the earlier reports are anecdotal with little or no quantification as to the levels of oil contamination: the actual types of oil stranded in the forests; the areal extent of mangroves oiled, and actual numbers, sizes, or areal ceverage of mangroves killed or stressed by the oil spill.

In addition, the species of mangroves involved was often not mentioned. In addition to impacts to the tress themselves, the funal component of the ecosystem is usually treated by noting dead or stressed animals at the time of the spill, or attempting to quantify losses or changes by one-shot sampling.

In particular, the sampling of infauna, in many of these cases, has not taken into account the normal seasonal variations in infunal species and numbers. All these problems may lead to widely different observations of the “impact” of a spill by different investigators.

Due to the fact that floating oil is transported by waves

and currents and strands on shorelines, low-wave energy ecosystems such as mangroves forests are routinely sites where oil accumulates after a spill. In addition, the inaccessibility of most mangrove forests makes oil removal very difficult, if not impossible, and the fine-grained and peaty anaerobic sediments characteristic of mangrove forests may reduce the rate of microbial breakdown of oil.

These factors, plus the burrowing activities of crustaceans characteristic of mangrove forests, can lead to persistent, high levels of oil contamination not only on the soil's surface but also deep in the sediments in the mangroves root core, and mangroves can be killed.

Table elsewhere in this chapter lists generalized responses of mangroves to oil spills in two general categories -acute and chronic exposure would be short term (1.4 weeks) and may and with no further impact if the oil is removed either mechanically or by natural have and current activity.

Chapter 7

MARINE OIL POLLUTION

Pollutants originating from domestic and industrial sources and from agricultural operations due to high inputs of *heavy metals*, *hydrocarbons* (*crude oil*), *pesticides fungicides*, *insecticides* have spread to coastal areas through rivers run off, via the atmosphere, direct discharges or dumping into the sea where the most of the active primary and secondary production direct, are being *polluted*, casusing serious damage to acquatic *ecosystems*.

OIL SPAS

Oil spill on ocean waters kills birds, fouls beaches, and poisons marine life. Hasan Jawaid Khan gives an account of these deadly oil spills, the damage they bring about and the ways by which an oil slick is contained tense of thousand of sea birds coated wtih oil shivering and dying on the beaches.

At least a thousand sea otters succumbing in the oil with their livers kidneys fatally impaired. And an unknown number of seals, whales and propoises poisoned.

Even before the tragic images of the deadly oil spill on toe Alaskan coast could fade from memory, the specture of an oil spill is back again to haunt us-only this time it is too bigger.

The enormous oil slick in the Gulf region, 30 times more than the Alaskan episod, threatens to set into motion an *unprecedented* ecological disaster that may take decades to fix. *Ironically*, the oil that drivers millions of vehicles round the world, sometimes drives countless *marine* animals to a most cruel death, crude oil when refined drives *automobiles* and airplanes, runs factories and farm *equipments*, provides gas for heating and cooking and is a source for *making drugs*, *cosmetics* and *fertilizers*.

But when crude oil spill on to ocean waters it spells death for marine animals and disaster for the *ocean ecology*. There has always been a danger of major oil spills Accui ring near a coastal area with *supertankers* beidg incrensingly used to carry *industrial* and *heating* oils.

Several of these huge ships have been grounded and cracked open by the seas, their vast *cargoes spilling* out heavy oil into the oceans where currents can carry it many miles before it disperses and *gradually* sinks to the floor.

Sometimes the currents carry the oil toward land where it fouls beaches and kills marine life. The world first woke up to the disastrous consequences of an oil spill when on 18 March 1967 a Liberian tanker Terry *Canyon*, ran a ground on the southwest cast of Great Britain, near the entrance to the English Channel, spilling 60,000 tons crude oil into the sea.

Oil splittered on to 160 kilometers of coastline killing countless number of fish and birds. Two years later, in January 1969 occurred the second major oil spill off the coast of Santa Barbara in the United States when an offshore oil well below out resulting in the discharge of oil at the rate of 1,000 gallons per hour causing extensive damage to to the coast.

In 1978, in the much publicized Amoco Cadiz disaster 68 million gallons of oil was dumped all along the French coast. And then on March 24, 1989 the supertanker Exxon Valdez belonging to the Exxon corporation, the largest petroleum company in the world, plied on to a reef off the coasts of Alaska.

Through five gashed in its hull the 30,000 ton super,

tanker spilled over 11 million gallons of oil into the clean water of Alaska's Prince William Sound. As the oil hit 1,930 kilometers of shoreline. 100,000 seabirds died including 150 raje species of haldeagles.

An unknown number of dead seals sank to the bottom of the ocean. And at least 1,000 sea otters perished. Even a few deer and a couple of bears that fed near the soreline were found dead. Closer home, Indian coasts have also been withness to a few tanker *disasters*, none too serious thought.

In July 1973, 3,000 tons of oil washed on to the Gujarat coast when an oil tanker cosmos pioneer, ran around. In 1974 an American oil tanker Transhuron colided with one of the atolls of the Lacadives spilling 5,000 tons of special furnance oil causing extensive damage to marine life.

Similiary destruction was witnessed when the oil tanker Lajpat spilled thousands of tons of oil along the *Bombayicoast*. The last time an oil spill *threatened* the India coast was in June 1989 when a Maltese tanker 1 M.T. Puppy collided with a British vessel.

The Maltese tanker spilled over 5,500 tons of furnance oil into the open seas off Bombay. At that time the fear of *contamination* of fish brought the fish industry to virtual halt at people stopped buying fish for fear it might cause cancer.

But not all the petroleum that gets into the marine environment is through oil spills. Natural seeps have been discharging petroleum into the marine environment for millions of years in amounts *substantially* greater than those resulting from present offshore production activities. More than 200 submarine oil seeps have been identified *worldwide*.

As assessment of petroleum pollution of the world's oceans by the national Academy of Sciences, Washington, estimated that between 1.7 and 8.8 million metric tons of oll enter the oce: ns annually. Oil spills are by far the major contributors to this pollution.

While other categories of pollution have shown a fifty per cent decrease, oil spills have been on the rise. But what is it that makes crude oil shore so deadly ? Crude oll is rarely used in the form of product at the offshore wells but it is converted in refineries into a wide range of products such as *gasoline*, *kerosene*, *diesel fuel*, *jet fuel*, *domestic* and industrial *fuel oils* and, *petrochemicals feedstocks.*

Until the nil is refined it contains potentially dangerous components, Crude oil is primarily made up of compounds of carbon and hydrogen ealled hydrocarbons. These *hydrocarbons* may be paraffins *parafin oil* is used as a fuel in *heaters* and *lamps cycloparaffins* (*naphthenes*) and aromatic compounds in varying proportions.

Crudes founds in eastei and aidwestern sections of the United states are predeminantly paraffinic, while those found along the Gulf coast are usually naphthnic. Crude oil also contains varying amounts of sulphur compounds, a small amount of nitrogen and very little oxygen or olefins which are unsatarated *hydrocarbons.*

Nickel and vanadium are present in all crudes in high concentration. Iron may be found in inorganic form as a result of corrosion of pipes and containers used to produce and transport oil. *Paraffins* like *methane* and ethane are asphyxiants, that is, cause suffocation. Some *paraffins* are central nervous system *depressants.*

Liquid paraffins can remove oil from exposed skin and cause dermatitis or pheumonia in lung tissue. However, these are the least toxic of the hydrocarbons. The effects of cycloparaffins are much like those of the paraffins except that unsaturated cycloparaffins are more noious than the saturated ones, breathing high concentrations of their vapour can result in Irritation and anaesthesia, the most toxic of the hydrocarbons are the aromatic hydrocarbons.

Inhalation of the vapour can cause acute intoxication. Benzone is particularly toxic and long-term expossure can cause naemia and leukopenia even as low concentrations. Sulphur compounds present in crude oil can also be toxic.

Carbonyl sulfide is dangerously poisonous. It is lethal to rates at 2900 ppm (parts per million).

The mechanism of toxic action appears to involve its breakdown to hydrogen sulfide. It acts principally on the central nervous system, with death resulting mainly from respiratory paralysis. Sulphur in the form of aromatic thiophenes, benzothiophenes, can be mo erately toxic. These damage the livers and kidneys.

Certain sulphur compounds like sercaptan can also be lethal. Although the long-term effects of an oil spill are not so well known, the Alaskan oil spill provided an opportunity to study the short-term effects As the 11 million gallons of oil spilled by the Exxon Valdex hit the shores, seabirds started shivering and dying.

The oil had broken down the natural insulating oils and waxes which shield the birds from water. As a result hypothermia (abnormally low body temperature) set in and the birds started dying. About 150 rate species of bold eagles also became victims when they ingested oil, scaven giving on oily seabird carcasses.

Cleaning away the oil from the birds has not been very successful as it causes a severe strain on the birds. However, the British Wildlife Rescue Association has designed wool sweaters which absorb oil from the strciken birds. The sweaters have already bean used to clean birds in the North sea and were found to be more effectively than cleaningwith detergents.

About 1,000 sea others died after their far became saturated with oil and provided no insulation. Others developed respiratory ailments as volatile components of the oil weakened membranes in their lungs. Some suffered liver and kidney damage caused by ingesting oil while cleaning their coasts.

Firstly were found to be the least affected. When fish from the oil spillage sites were tested no hydrocarbons were found in their flesh. The fish are efficient at converting hydrocarbons they ingest into metabolities which are then excreted from the liver to the gall bladder. Moreover, because of their mobility the fish escape large-scale contaminated as they are able to avoid heavily contaminated areas.

When oil spills on to ocean waters it spreads rapidly over the surface. The force of gravity causes the ligher oil to seek a constant level by spreading horizontally on the heavier water. The movement of the oil is caused by the combined action of wind, surface currents waves and tides.

Oil on water normally moves at about 1/30th of the wind speed. With large amounts of oil, as with accidental spills, the oil drift is largely independent of spill volume, spreading, or weathering. Since the thick portion of the slick drifts faster than the thin portion a heavy oil accumulation forms the leading edge of an advancing slick.

As the oil spreads it comes under the influence of several weathering processes that ultimately break down a spill with time. The first of these is evaporation which is dominant soon after oil is released. Evaporation is determined by wind velocity, nature of the roughness, and temperature.

Almost 50 per cent of oil is lost through evaporation during an oil silk's lifetime. Some of the light, low-bowling hydrocarbons, such as benzene, toluence, and xylene are lost through evaporation. Their removal decreases toxicity of the oil to marine life. Much of the oil that evaporates is photooxized in the atmosphere and some of it may return to the seats as atmospheric fallout.

A very small fraction of the oil dissolves in the water in the early stages of an oil spill. Lower-molecular-weight compounds tend to be the most soluble. However, this process is very insignificant. Natural dispersion also removes some oil from the water. In this process, small droplets of oil which are larger than dissolved molecules, are incorporated into the water in the form of dilute oil-in-water suspension.

Crude oil contains trace amounts of altrogen, sulfur, and oxygen-bearing compounds which can act as natural surfactants. These surfactants reduce the oil-water

Table 7.1: Mineral Medium.

Salts	*Concentration (g/L)*
Sodium Chloride	23.9
Megnesium Chloride	10.8
Calcium Chloride	1.15
Potassium Chloride	0.74
Sodium Sulphate	0.40
Sodium Sulphate	0.023
Sodium Bromide	0.117
Sodium fluoride	0.002
Sodium Bicorbonate	0.196
Silican Dioxide	0.022
Sodium Oxide	0.022
Ferrous Sulphate	0.002
Amonium Chloride	0.9
Potassium Hydrogen Phosphate	0.15

interfacial tension allowing the oil to break up and to disperse into droplets more readily.

A part of the oil is also broken down by micobial degradation. Microorganisms present in the water have a great capacity to utilize hydrocarbons present in the oil as in energy source. This process especially speeds up when she oil disperses.

Some organisms may also ingest dispersed oil droplets and subsequently deposite them as fecal pollets. Emulsification is yet another weathering process in which water is incorportated into the floating oil forming a water-inoil emulsion.

Such emulsions, which may contain from 20 to 80 per cent water, are often very viscous and referred to as "*mousse*" once mousse formation starts, cleanup operation run into difficulties as it is almost impossible to pump. Ultimately the mouse is broken up by wind and wave action into tar balls that may wash ashore.

However, breaking up of an oil spill takes time. And the longer the oil continues to float on water the more damage it does to the marine environment therefore, cleaning no an oil spill fast *asumes* prime importance.

An oil spill can be cleaned up by mechanical removal of the oil, by the use of *chemical* and *microbial* surfactants, by using absorbants, or by burning the oil slick.

Spills can be dealt with more easily if they are confined to a small area on water surface. For this, mechanical booms or barriers are spread around on oil slick to check its progress and prevent it form hitting the shoreline. But sometimes high waves may *splash* the oil over the *booms*.

The oil may even pass below the booms once it has piled against the barrier. So after the oil has been contained by the booms it is sucked up from the surface of the sea by *vessel* known as *skimmers*. Skimmers now available generally fall into one of two categories.

The first, mechanical surface skimmers, remove the top layer of the water and oil from the surface but they are not very *effective* if there are high waves. They *gulp* large amounts of water unless some provision is made to allow the weir suction *port* to follow the water surface.

A second type of skimmer operates on the principle of selective wetting of a surface by oil rather than by water. *Rotating metal* disk or *conveyor* belts dip into the water, a *surface* layer of oil removed.

The spilled oil can also be treated with dispersants. These chemicals are sprayed from aircrafts or ships. *Dispersants* cause the oil to spread further and disperse in a manner similar to the way soap removes oil from one's hands allowing the oil to be emulsified and washed away with water dispersant contains a surfactant, a solvent, and a stabliizer.

The solvent usually comprises the bulk of the sispersans and enables the surface-active agent or surfactant to mix with, and penetrate into the oil *slick* and form an emulsion. The *stabilizer* fixes the emulsion and prevents it 'from coalescing once it is formed. *Dispersion* increase the slick *suaface* area and allows a rapid increase in the rate or *microbial* decomposition.

But although dispersants are effective in diffusing a slick, these chemicals are toxic to *marine life*. And since the emulsified out droplets sink into the water body, oil stays in the waters with its toxic components such as benzene and *toluence* entering the food chain with disastrous 1ongterm effects.

Absorbents are also used to facilitate the cleaning up of oil spilts. When they are applied to the slick, they abosrb the oil and prevent it from spreading, and when the absorbent material is removed from the water, the oil is removed. Natural materials such as peat moss, straw, sawdust, and pine bark can be used.

Synthetic absorbents include polyethylene, *polystyrene*, *polypropylene*, and *polyurethane*. Of ail *synthetic absorbents*, *polyurethane* is the mot promising. Burning oil slicks on the open seas has generally been less successful because the more *volatile light* fractions evaporate quickly from the oil slick.

Also, the water removes heat fast than it can be created to support the combustion. But burning oil slicks also leads to extensive air pollution. By far the safest way of treating an oil slick is by using biological agents for degrading oil. These *microbial surfactants* are sprayed from the air.

They mix with the oil, emulsify it and disperse it throughout the water body thinly that it no longer remains *hazardous*. However, not very cleau-up method can clear up all oil spill. Clean-up methods are chosen on a case by methods are chosen on a case-by-case basis.

But whatever the method, it is important that thc cfforts are mounted as soons as the oil spills is discovered. For the longer it takes for the cleaning up to start, the more distance the oil travels. And as the oil charts its destructive course it fouls beaches, kills marine animals and *contaminates* the food chain.

Gulf Oil Slick Making History

According to H.J.K. the massive oil slick in the Gulf is making history. With the total spillage of oil in the Gulf waters now standing at more than 330 million gallons, it is already 30 times more than the quantity spilled by the Exxon Valdez on the shore of Alasks.

The oil slick that now stretches over an area more than 80 km long and and 20 km wide is moving south at a speed of 20 km a day. Having already hit the Saudi shares, affecting birds and marime life, the slick is now threatening the shores of Bahrain, Gatar and the United Arab Emirates.

Apart from killing sarine life the oil also theatens to disrupt the deslination plants on the western coast of the Gulf that provide nearly half the wates to Arabian states. The enormous oil slick will disrupt the already fragile ecosystem of the region.

According to the world conservation *Monitoring* Centre in Cambridge, between 1 million and 2 million birds will die as a result of the gulf oil spillage, Some species may never recover. Cormorants coates with oil have already started dying on the Saudi beaches.

Dolphins may escape *contamination* as they seem to be able to sense a spill and streer clear of it. The *dugong* or sea-cows will be affected with their food supply of sea or grass getting contaminated. The three already endangered species of turtles the green, leatherback and hawks bill may be *seriously* affected. *Coral reefs* housing a di,rerse variety of fish may be *destroyed.*

The oil is likely to affect the *mangroves* of Iran and a number of valuable shrimp fisheries. No *immediate* threat to the Indian coastline is perceived so far according to oil experts The oil spilt from the Al-Ahmadi terminal is *export* quality *Kuwaiti crude*, extremely light and fluid. Hence, much of it would evaporate.

In the first 24 hours, 35 per cent of the crude is likely to evaporate and in the next 10 days about 45 per cent. Even if some of the oil manages to find its way out of the

Gulf, much of it would be broken up by the rough tropical waters of the Arabian Sea.

Due to burning of the oil in the Gulf and attempts for its removal, a very small quanity of tar balls may eventually hit the Indian coast. These tar palls can just be collected and set on fire.

Gulf Spill: No Threat to India

Biman Basu made survey on Gulf oil spill and he pointed out that the oil spill in the Gulf is the largest so far, spreading over 700 sq. km, it does not pose any *pollution* threat to Indian waters. According to Dr. Peter.

"This is not so much environmental *terrorism* as an environmental crime." said Worldwatch founder Les Brown, "We need a code of ethics like the Geneva Convention (governing treatment of prisoner of war) for the *environmental field.* The Gulf could die in a matter of months."

Ancient warriors knew how to use the enviroment as a weapon. Corpses were catapulted into besigned cities to spread disease; water supplies were *diverted* to cause drough or floods; wells were poisoned.

The Romans spread salt on the fields of Cartlage to deny food to their enemies; *Russian* troops defeated *Napoleon's* army by torching the earth as they retreated, *destroying* all crops animals and human dwellings.

Prime Minister Winston Churchill planned lo spread burning oil on Britain's *beaches* to block any German *invasion* in the Second World War. In the same war the Dutch opened up their sex dikes to food their own land in a bid to stop the advancing *Nazi troops.*

The U.S Forces in Vietnam also *practised* environmental warfare when they sprayed Agent Orange to defoliate waste *stretches* of forest land deprive Viet Cong *guerillas* of *protective* cover.

The spraying destroyed large number of animals that lived in the forests, polluted waterways with the reputedly,

Table 7.2: N-Alkanes Composition of Saturated Hydrocarbons.

C-Number	*Arabian Light 240+*	*C-Number*	*Arabian Light 240+*
15	3.44	26	3.75
16	7.72	27	3.18
17	10.13	28	2.54
18	10.42	29	2.05
19	9.73	30	1.85
20	9.10	31	1.47
21	8.23	32	1.20
22	7.03	33	1.01
23	6.12	34	0.76
24	5.37	35	0.52

cancer-causing chemical dioxin, and Vietnamese exposed to *defoliated* areas reported *increased* rates of miscarriages and birth deformities. *Heydemann*, Science *Counsellor* in the US Embassy in New Delhi, the oil may not come out of the Gulf for years because of the current patterns in the region.

By that time most of it would have been *disintegrated* by natural bac teria. But the slick will cause irresparable damage to the rich marine ecosystem of the Gulf.

The Environment as a Weapon

Ben Barber studied that the paint oil slick released into the waters of the Gulf during the war has been denounced as "environmental terrorism" with along with *deliberately* set oil well and refinery fires, could after the region's ecology for the years to come.

Yet, since Biblical times, armies have destroyed the environment to deny and enemy food, water or shelter. "The *phenomenon* is not new there have been efforts to degrade the environment as part of war fare for sometime," said Mr. Chris Flavin an analyst with the Washington

Table 7.3: Hydrocarbons Analysis

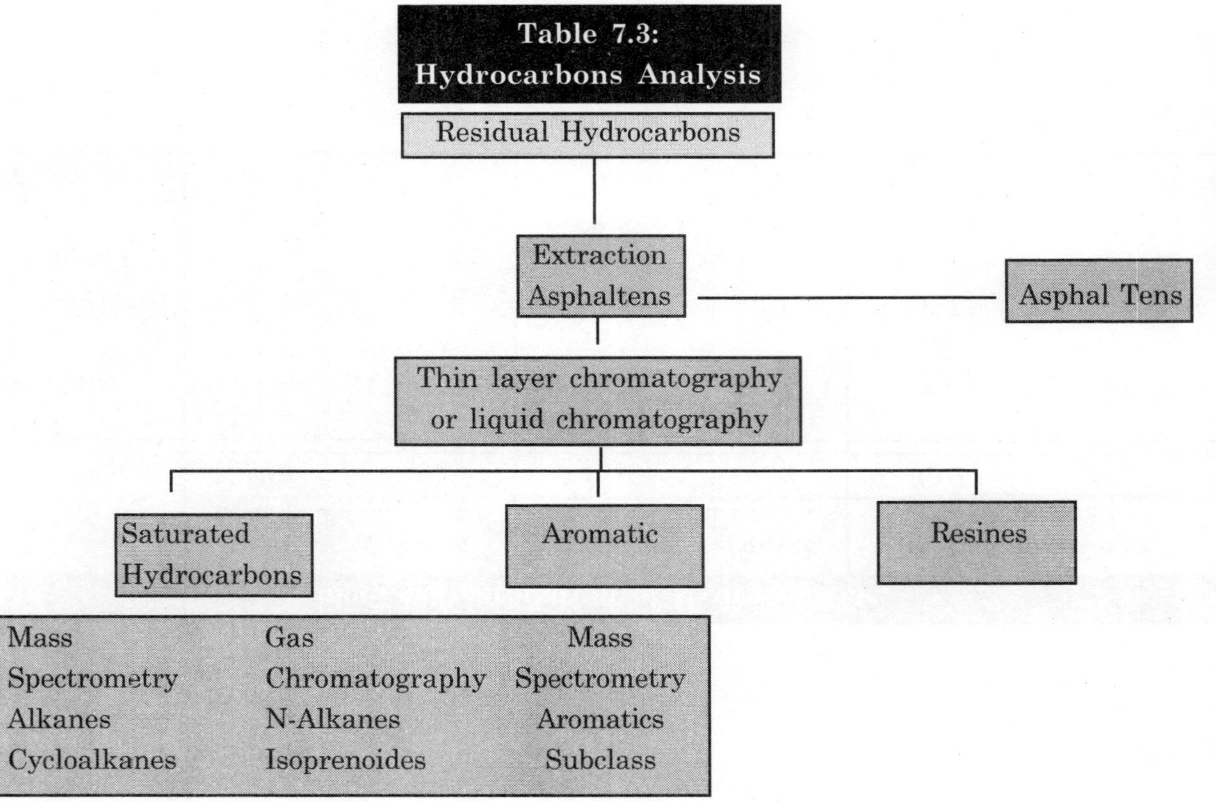

Table 7.4: Degradation of Hydrocarbons.

	Start of batch		*End of the batch*		*Hydrocarbons consumed*	
	[g/L]	*%*	*[gfL]*	*%*	*[g/L]*	*%*
Total Hydrocarbons	2.65	—	1.57	—	1.08	41
Saturated Hydrocarbons	1.17	44.2	0.39	24.9	0.78	67
Aromatics	1.00	37.5	0.73	46.5	0.27	27
Resins	0.42	15. 9	0.39	24.2	0.03	7
Asphaltenes	0.06	2.4	0.06	4.4	0.00	0

Table 7.5: Degradation of Saturated Hydrocarbons.

Hydrocarbons	*Start of batch [mg/L]*	*End of the batch [mg/L]*	*Hydrocarbons consumed [mg/L]*	*%*
Saturated Hydrocarbons	1170	390	780	67
Alkanes	739	82	657	88.9
1—ring	99	51	44	44.4
2—ring	136	89	47	34.5
3—ring	92	79	13	14.1
4—ring	55	50	(5)	—
5—ring	27	24	(3)	—
6—ring	22	15	(7)	—

based environmental *think-tank*, *Worldwatch* institute. But modern technology has upped the ante." he added "There's a much greater impact."

The estimated 460 million gallons of oil released in to the shallow a waters of the Gulf is by for the world's greatest oil spill. Spreading south from off the *Kuwaiti* and *Saudi* coasts towards *Bahrain* and *Gatar*, it threatens millions of sea birds and marine life forms, including the rare dugong or sea cow, the *mammal* that *supposedly* gave rise to the *legend* of the *mermaid.*

Since the mar ended in 1975, low-value bamboo and grasses have invaded defoliated areas, preventing the regrowth of valuable hardwood forests.

The U.S. also seeded *clouds* over Asia to try to alter Vietnam's rainfall and make supply trains to muddy to use, said *Prof. George Ratbjens* of the *Massachusetts* Institute of Technology.

A political science teacher who has specialised in nuclear weapons and *environmental* issues, he said some modern *military* planner have *proposed* cloud seeding to divert rainfall from an *enemy's crops.* In addition to the releases of oil into the Gulf, several *Kuwaiti oil* refineries were set on fire in what may have been an attempt to raise a smoke shield against attacking *aircraft*, Some scientists have warned that such huge fires could block out *sufficients* sunlight to disrupt monsoon rains as far away as South Asia.

The worst-case scenario warned that massive famine could result. Prof. Rathjens, however said it is unlikely the fires would result in the phenomenon known as "*Nuclear Winter*" the predicated effect on the planet of an *atomic war*.

Biodegradation of Crude Oil in a Marine Environment

Ballerini-D and J.P. *Vandecasteele* studied the aims essentially at quantifying the processes of *microbial* degradation of crude oil in optimal conditions compatible with a marine environment.

Our purpose, specifically, was to determine the, porportion of crude which could undergo biological degradation, and to improve our knowledge of the biodegradation kinetics of the various *hydrocarbon* families which *constitute* a *crude oil.*

This work was performed in laboratory reactors where various *physiochemical parameters* such as pH. temperature, stirring speed and air flow rate were strictly controlled. The study was conducted first in batch then in *continuous cultures*.

The composition of the mineral medium used for the culture is shown in Table elsewhere in this chapter. This composition was chosen by taking the average composition of the *Atlantic Ocean* as a basis (1) and entriching it in *nitrogen*, *phosphorus* and *iron*.

The *concentrations* of these three elements were respectively raised from 0.5 to 235 mg/L for nitrogen, from 0.007 to 26.7 mg/L for phosphorus and from 0/001 to 0.4 mg/L for iron and the pH of this medium was adjusted to

8.1 the average pH of seawater.

In order to approximate the conditions existing at sea, where evaporation is not negligible (2) we mainly used crude oil which had been topped at 240 degree C. In the present case, we used a residue of an *Arabian light crude*, obtained by *distillation* under reduced pressure, in which all the factions *distilling* below 240 degree C had been removed.

Table elsewhere in this chapter shows that in the alkane distribution of Arabian light crude oil topped at 240 degree C *n-paraffins* of chain length below C 14 have been completely removed. In oil spills at sea, all compounds with a number of carbons below 15 disappear after a residence time of about 10 days.

The reactor used in our studies is glass cylinder equipped with two stainless steel end plates. Its total volume is approximately 3 liters. Stirring is achieved with a simple 4-bladed turbine, the speed of which is adjusted between 600 and 800 rpm.

The reactor is equipped with automatic pH of the culture is maintained at 8.1 by addition of 1 N *sodium hydroxide*, the temperature is kept at 20 degree C=0.2 degree C and the air flow rate is maintained at a value of 1 liter air/L of medium/hr. Experiments are conducted in nonsterile conditions.

The analytical methodology utilized to follow the processes of crude oil degradation is also *summarized. Gaseous* elluent from the f'.rmentor first goes through a flask containing carbon *tetrachloride* where the hydro carbon vapours are trapped, then in a second flask containing a known amount of IN potassium hydroxide. which traps carbon dioxide.

Hydrocarbons are then determined by infra-red spectrophotometiy and the carbon dioxide is determined by titration. Separation of asphalotenes is first performed on the hydrocarbon residue. Asphaltenes are precipitated by boiling with heptane under reflux for 1 hour.

Insoluble materials are then separated by filtration, rinsed with hot heptane, then dried and weighted. On the residue obtained evaporation, separation of the three main classes of compounds of crude oil, saturated hydrocarbons, aromatic hydrocarbons and resins is performed either by thin layer chromatography on silica gel or by liquid chromatography on a silica gel column using cyclohexane as the elution solvent.

The sum of the weights of the four fractions recovered in this procedure always represents 90 to 1(10% of the initial weight of the total oil sample. On the saturated and aromatic fractions, more detailed analyses using gas chromatography and mass spectrometry are then performed.

The Isolation and selection of the strains of microorganisms used were performed starting from marine silt and sand samples collected at places hit by crude oil spills.

The first enrichment cultures were performed by successive transfers in liquied medium to which 2 g/L of crude had been added, first in shaken flasks and, soon afterward, in laboratory fermentors, in order to obtain perfectly defiled culture conditions. In the course of these successive batch cultures we observed a very important decrease of the lag phase and an improvement of the final biodegradation performance.

A large improvement of the dispersion of the hydrocarbon phase in the mineral medium was simultaneously observed. A very fine emulsion of the organic phase in the aqueous phase was thus obtained, leading to a marked increase of the reactional area.

The bacterial association present in our selected cultures consists of four dominant bacterial genera: Pseudomodas, Moraxella, Acinetobacter and Flavobacterium. During a batch culture we studied the changes in the bacterial flora. Although at the beginning the dominant strains belonged mainly to the genus Moraxella, the Pseudomonas strains predominated later on, as oil biodegradation advanced.

The biodegradation process during the batch cultures can be illustrated by following the changes of various parameters. For ex ample, we have represented biomass and carbon dioxide production, the latter in both the aqueous and gaseous phases. Total production of Carbon dioxide follows an evoluation quite comparable to that of biomass production.

At the end of the culture, whereas hydrocarbon consumption reaches 2 g/L the total amount of carbon dioxide produced is 1.45 g/L and that of biomass is 1.4 g/L. The changes in time of dissolved oxygen concentration in the medium, as well as consumption of ammonia nitrogen and hydrocarbons are shown in figure elsewhere in this chapter.

The amount of hydrocarbons lost by evaporation is quitenegligible as it is measure on a 48 hour period and is always below 10 mg. A perfect correspondence between the

Table 7.6: Degradation of Aromatic Hydrocarbons.

Hydrocarbons	*Start of batch [mg/L]*	*End of the batch [mg/L]*	*Hydrocarbons consumed [mg/L]*	*%*
Monoaromatics	268	138	130	48.5
Diaromatics	251	168	83	33.0
Othercompounds	481	424	57	11.8

Table 7.7: Degradation of Monoramations.

Hydrocarbons	*Start of batch [mg/L]*	*End of the batch [mg/L]*	*Hydrocarbons consumed [mg/L]*	*%*
Alkylbenzenes	113.4	36.6	76.8	67.7
Benzocyclo Paraffins	100.0	53.8	46.2	46.2
Benzodicyclo Paraffins	54.9	47.4	7.5	13.7

evolutionof the consumption of ammonia nitrogen and hydrocarbon can be again noted.

In addition, during the acceleration period of the biodegradation process, a very distinct decrease in the oxygen dissolved in the medium occurs.

Actually, dissolved oxygen concentration even becomes a limiting factor since values of 5% of the saturation level are read. Later, **as** the velocity of the bio-degradation process decreases, the concentration of oxygendis3olved in the medium increases again, but without reachingthe saturation value.

The results of mass spectrometric analysis of the saturated fraction are presented in Table elsewhere in this chapter. First they indicate that alkanes are the main target of biodegradation since 88.9% have disappeared at the end of the culture.

This allows us to conclude that, in addition to n-alkanes and isoprenoids which represent only 14.8% of the saturated fraction, the great majority of isoalknaes is consumed by microorganisms. Among naphthenic compounds, 1-ring and 2-ring naphthenes are the main targets of biodegradation, with respective percentages of biodegradation reaching 44 and 47%. For the the aromatic fraction results presented in Table 11.5 indicate that the microbial action is particularly evident at the level of mono-and diaromatic compounds.

At the end of the culture all mono and diarorratic compounds with a carbon number below 16 hive disappeared. Among monoaromatic compounds, it can be seen in Table elsewhere in this chapter that the compounds most susceptible to microbial action are alkylbenzenes, 67.7% of which have disappeared at the end of the culture, and benzocyclop-arafiins, 46 2% of which have been consumed during this experiment.

The results presented in Table elsewhere in this chapter illustrate the microbial action on diaromatic compounds and, in particular, the effect of microbial attack on the residual content in naphthalene derivatives, 50% of which are

Table 7.8: Degradation of Diaromatic.

Hydrocarbons	*Start of batch [mg/L]*	*End of the batch [mg/L]*	*Hydrocarbons consumed [mg/L]*	*%*
Napht halenes	98.3	49.1	49.2	50.0
Acenaphtenes	81.4	60.4	21.0	25.8
Fluorenes	71.0	58.3	12.7	17.9

consumed during the biodegradation process. The primary purpose of this paragraph is to provide some guidelines for rehabilitating coral reef habitats that have been impacted by oill spills.

The rehabilitation process inthe final step of a spill response effort that also includes actions to contain, cleanup, assess damages, and miltigate damages from a spill.

Because our ability to rehabiliate damaged coral reefs is limited, this ordering of events also represents priorities for actions to be taken when a reef is threatened by oil pollution. In view of these priorities, the objectives of this chapter are threefold

— To identify reef areas particularly sensitive to oil pollution incidens and to suggest means to minimize the threat of such incidents.

— To suggest techniques for assessing the impact of oil pollution incidents on coral reefs and their associated communities, and

— To provide guidelines for rehabilitating a reef after it has been impacted by oil pollution.

CAROL REEFS AND PETROLEUM INFRASTRUCTURE

Coral reefs are structures created and maintained by the establishment and growth of populations of hermatypic corals and coralline algae. They may be comprised of wave resistant emergent of submerged reefal zones, or a combination of both.

Geomorphic ally barrier and fringing reef protect the insullar and continental coastlines ttat they border from erosion. Reef rock that forms terrestrial substratum on islands and recently uplifted coastal margins is often the sole or major component of limestone.

Geologically older reef formations are found far inland and may serve as porous, oil bearing strata. Reefs typically provide habitat for a large variety of attached plants and epifauna, infauna mobile invertebrates, and fishes. The large numbers of economically important species they support make reefs locally important to commercial sports fisheries.

The resultant high diversity and abundance of reef associates and the functional and spatial dominance of corals and coralline algae are the essential characteristics of coral reefs. Coral reefs are circumglobal in the tropics and subtropics between the norther and souther hemispheric 18.5 degree climatic isotherms.

A majority of coral species and the most diverse and prolific reefs occur in Indo-West Pacific seas. Tropical West Atlantic and Eastern Pacific resfs are generally less diverse in terms of caroals and reef associates, but are functionally quite as important.

In a recent interview of the impact of oil upon corals, Ray (1980) pointed out that extensive tanker traffic and oil-related activity occurs within the biographic zones occupied by coral reefs. Sufficient information exists to enable the production of detailed maps showing the location of coral reefs in relation to petroleum infrastructure.

Such maps may be a useful aid in the identification of high risk areas. Two areas that have well developed reefs and much petroleum activity are given as examples in this discussion.

BIOLOGICAL EFFECTS ACTUALLY OBSERVED

Having now made the effort to consider from theoretical point of view the system which it is thought may give rise to biological effects, it is not time to consider whether these effects can actually be observed under the conditions which prevail in the marine environment.

Natural Fluctuation

The marine environment is, however, subject to a wide range of influences the effects of which may swamp any effects which may be due to pollution. This is the well-known signal/noise ratio problem commonly encountered in the physical sciences.

On the other hand, if studies are conducted in the laboratory the results can be no more than indicators of what will happen in the real world because it is quite impossible to reproduce the marine environment in the laboratory.

Thus on the one hand we have the problem that the data obtained in what may be well-regulated laboratory experiments, in which the pollution parameters may be isolated for study, do not give results necessarily applicable to the marine environment, and on the other we have real environment studies in which the pollution effects are not isolatable and are usually swamped by othermore influential effects.

Correlation of Results Obtained in the Laboratory and at Sea

The review also covers results obtained by exposing organisms to oil in the laboratory. Here again there are difficulties. Often the concentrations used in order to observe an effect are higher than could possibly be experienced in the marine environment itself or are continued for greater periods of time. Again, even when the concentrations are comparable, it is impossible to extrapolate to the real situation.

It is quite clear that one can measure in the laboratory lethal and sub-lethal effects on organisms exposed to dissolved oil components and disoersed droplets of oil in the water column. It is also clear that the concentrations of oil in the water column at real spills can be broadly comparable to those in the laboratory studies. In addition similar effects, both lethal and sub-lethal, on planktonic species can be observed at sea under real spills.

When one attempts to evaluate the significaupe of those observations, however, in the only manner possible or meaningful, *viz.* by attempting to observe subsequent changes in fish populations no such changes are discernible.

There are a number of quite reasonable explanations for this slate of affairs. Thus even if plankton populations are reduced significantly under oil slicks, phytoplankton is in continuous production and the is rapidly restored as the oll alick moves on.

Even if fish egg and larval stage numbers were similarly

Table 7.9: Chemical dispersion of 1 mm thick slicks of Kuwait Oil.

Time after discharge (min)	*Concentration of Kuwait crude oil (ppm) in upper metre of sea water*		
	Run 1	***Run 2***	***Run 3***
0	34.4	24.2	15.8
1	—	15.8	—
2	47.8	—	8.7
2.5	—	12.2	—
5	—	9.4	—
7	17.8	—	3.5
10	—	5.2	—
15	—	—	1.7
18	1.9	—	—
25	—	4.2	—
40	0.8	—	1.35
50	—	1.9	—
80	—	—	1.5
100	2.2	0.8	—

reduced it is fact that very few would have reached maturity anyway. The losses caused by the oil merely cause losses which would otherwise have taken place naturally.

Larger Marine Organisms

The conculusion reached above regarding plankton in general do not of course rule out hte possibility that larger marine organisms may be affected. When this possibility is considered, however, it is found that free-moving fish and crustacea are largely unaffected.

Sedentary crestacea on the other hand may be subject to smothering with an oil layer deposited by the outgoing tide.

Grazing molluses and. of course, sea-weeds are liable to similar fate while filter-feeding molluscs can ingest oil in separating out food particles.

Generally mortality is highest with fresh oil, weathered oil having little effect. In any case the effects are transient as regards populations and deputation occurs rapidly with filter feeders when exposed to clean water.

Some species are not sufficiently mobile to escape, do not posses shells by which to isolate themselves or have other protectiue coatings.

Worms living in sandy or muddy beaches may be killed if oil components penetrate the substratum. Of course, oil will not penetrate across the water-table interface so this will occur mainly in well-drained beaches.

Birds

With birds it can be concluded that soluble and sipersed oil components will have no effect and we have now reached a pure example of the effects of physical coating.

There is no doubt that this form of pollution has a very serious effect **at** least on individual birds, but here again the Royal Commission has not been able to demonstrate conclusively that there is any threat to populations.

There is, however, considerable cause for concern, particularly in relation to major spills close to a breeding colony onshore. It is obvious that oiled birds are one of the main vehicles of public awareness of oil pollution and arouse widespread feeling of pity and revulsion.

Large Mammals

There seems little possibility of affecting large marine mammals such as the cetaceans. Seals on the other hand may become oiled as they leave and re-enter the sea and this could result in loss of insulation.

It has been noted, however, that seal populations around the UK are generally stable and that grey seal numbers have been increasing dramatically for the past 50 years.

Human

In the early 1970s attention was given to the possibility that polyaromatic hydrocarbons present in oil pollution might concentrate in the tissues of marine organisms and that those concentrations might increase up the food web to levels which could induce cancer in man.

Research results now available, however, indicate that in common with other petroleum-derived hydrocarbons these compounds are rapidly released when the organisms are exposed to clean water and it is extremely unlikely that they could accumulate in the manner originaliy feared.

It has also been shown that thesc compouns are in any case the inevitable products of the combustion of most types of organic material such as *wood, garden rubbish* and even *tobacco*, and as such are widely distributed in the environment.

Furthermore they are present in many foods and it is therefore possible to show on the basis of dietary studies that the consumption of sea food would be unlikely to provide more than 2-3% of the normal intake.

Recent research has also failed to establish a connection between this form of pollution and cancer in the marine animals themselves and it is extremly unlikely, to say the least, that ingestion of cancerous tissue from animals could cause cancer in man in any case.

AIMS AND OBJECTIVE OF OIL SPILL RESPONSE

The aims and objective of oil spill response should therefore be to protect amenity, fishery and wildlife resources and the efforts expended should ideally have some relationship to the quantification of threat to those resources if nothing is done.

If nothing is done to respond to an oil slick far enough at sea for it to disperse through natural forces before it *reaches* shore, then minimum impact on the environment will result since there will be no amenity loss, and no effect on inshore fisheries. The oil will disperse far at sea when its effect is least.

The only possibility of *quantifiable* damage resulting would be if high concentrations of seabirds became affected. If natural dispersion forces are unable to remove the oil from the sea surface before it reaches shore, then, in addition to seabirds, inshore fisheries, the flora and fauna of the inshore zone and *amenity* are at risk.

In terms of '*quantified risk*, therefore, there is no likely to be greater environmental impact than in the previous example. It is reasonable therefore to make greater efforts to respond in this case than in the previous one.

It should however, be reasised that efforts should be made to deal with the oil in the lower-risk are, i.e., at sea, before it reaches the comparatively high-risk area, i.e., the inshore zone. If it is possible to prevent oil reaching the high-risk zone, minimum damage will result.

If response occurs only when the oil has reached the high-risk zone then the damage will already have occured and one is then no longer preventing damage but only attempting to assist the area to recover from the damage it has already sustained.

It is therefore clearly preferable to deal with the oil in lowrisk areas before it reaches those of high risk. If this is impossible, then the aim should be to return the damaged areas to their original state as quickly as possible and the best way to assist in this will be to remove the oil as quickly as possible if natural processes are considered to be too slow.

Methods of Oil Spill Response

There are three possible responses to oil spills at sea.

- It can be decided to have the oil to be dispersed by natural processes.
- The natural dispersion rate may be increased by the use of chemical.
- Attempts may be made to remove the oil from the sea surface by mechanical means.

The nil respose option will be possible when no resources are threatened and natural dispersion rates are fast enough to remove the oil from the surface before any resource comes under threat.

The use of chemical dispresants should be seen as a means to increase natural dispersion rates so that resources are less likely to come under threat.

It is of course preferable to remove the oil completely from the environment, and so mechanical recovery devices are favoured provided they will work sufficiently to deal with the oil in the low-risk zones before it reaches those of high risk.

It is generally accepted that if recovery is not possible far at sea in a given situation, it may be preferable to remove the oil completely from the environment, and so mechanical recovery devices are favoured provided they will work sufficiently to death with the oil in the low-risk zones before it reaches those of high risk. It is general accepted if recovery is not possible far at sea in a given situation, it may be preferable to disperse the oil there and so prevent damage later in a high-risk sheltered zone than to wait and attempt to recover it when it has reached the high-risk sheltered zone.

The latter option will not prevent damage. It will merely remove the oil after the damage is done. Such belted action will of course assist in the recovery of the area affected.

The Use of Dispersants

Clearly oil does not remain permanently on the sea surface. The pollutant may disperse into the water column either completely or partially under the action of natural forces and or it may come shore. If the oil is spilled close to shore there is little opportunity for natural dispersion before the slicks reach shore.

In such cicumstances beach pollution seems inevitable. However, even in the Amoco Cadiz it has been estimated that some 50% of the spillage dispersed into the water column in the prevailing conditions of high natural agitation rather than stranding on the beaches.

With the advent of North Sea oil exploration and production, efforts were made to estimate the quantitities likely to reach shore by considering natural dispersion rates in relation to the speed or travel on wind and tide.

It had been concluded that for Ekofish oil only very small percentages of the original spillages have any chance of polluting beaches. Thus although natural dispersion does take place, it cannot be relied upon in general to remove the pollutant from the sea surface before it reaches shore.

The extent to which natural disprrsion will assist is a function of oil type, quantity spilled and distance from shore of the spillage source. It is important to recognise at this point that dispersants are simply as means of increasing the natural dispersion rate so that sufficient before it reaches the showe.

Certainly it is necessary to treat an in which this is an advandage. Thus the spreading phenomenon itself is form of dispersion. In fact it is two-dimensional dispersion and all that remains to be done is to add the third dimension, i.e., to disperse the surface oil into the body of the sea.

Oil concentrations measured under chemically dispersed oil slicks and under the naturally dispersing Ekofish Bravo oil slicks were much lower than 100 ppm because the oil, of course, disperse, throughout the sea water to even greater depths than 1 m. Further experiments on this topic have been carried out on Kuwait crude oil at experimentally controlled thicknesses, with the fan-jet nozzle arrangement used for dispersant efficiency testing at sea.

Here slicks up to 1 mm in thickness were produced in order to produce high initial concentrations in the water by use of appropriately enhanced dispersant discharge rates. Maximum possible concentration in the top metre on dispersal of a 1 mm thick slick would be 1000 ppm as opposed to the 100 ppm arising from the normal 0.1 mm thick case.

The aim of these experiments was to study rates of dillution of the dispersed oil into the sea following treatment. Wind speeds on this occasion were 8-10 knots with sea state 2-3. With this degree of agitation, observed concentrations were low, suggesting rapid transportation of oil to greater depths by natural tubulence as shown in Table elsewhere in this chapter.

When these reults are compared with the theoretical concentration of 1000 ppm. if all the oil remained uniformly dispersed in the top metre, three is a strong suggestion that intial concentrations in surface waters are reduced by dillution to greater depths. The results of these chemical dispersion experiments in clam and slightly rougher conditions suggest that while initial concentrations might be higher than for natural dispersion, subsequent dillution is rapid, even for slick thicknesses ten times thicker than those normally treated.

As a result the persistencc of significant amounts of oil in the surface waters is likely to be of shorter duration if dispersants are used than if they are not.

The question of the possible impact of dispersed oilbn organisms has been taken further by McAuliffe and his coworkers, who have demonstrated that evaporation tends to ensure that the dispersed droplets do not include the toxic fractions of the crude oil.

Oil Recovery at Sea

Oil recovery at sea it is necessary to return to the earlier assessment of the problem. It will be recalled that the counter rate for a I m wide unit travelling at 1 m 0.2 tonnes/h.

It is also a known fact that if a barrier having draught and free board a skimmer or a boom for example is moved across a water surface at speeds greater than 1 knot there is set up a flow of water beneath the barrier which is sufficient to carry with it any floating oil which is in front of the barrier: consequently the encounter rate in the oil recovery context cannot be increased by increase in speed of travel in the manner appropriate to disperent application.

At first sight the only option is to increase encounter rate by increasing the width of the collection unit by the use of booms. Thus a boom array of width 250 m will increase encounter rate of 90 tonnes/ h. Such an array, however, is extremely cumbersome in practice.

To remove oil pollution from the marine environment before it could reach sensitive inshore areas and beaches and from a realisation that, in spite of claims to the contrary, no significant quantities of oil had ever been recovered at sea.

The objectives of the programme were to find out why performance was so low and how to improve it with due regard to costbenefit concepts.

To this end, available equipment was classified in terms of the nature of the skimming element, *e.g.* absorption belt, disc. rope vortex chamber of weir/direct suction device.

In similar manner booms were classified in terms of their identifiable design elements. Examples of each type of stimmer and boom were them acquires from evaluation at sea in the presence of oil.

As a result of these investigations it has been found that recovery equipment fairs to perform satisfactorily in waves for the following reasons

— The inertial mass of the equipment is too great.
— The linear dimensions of rigid elements, or of the entire systems, are too great in relation to encountered warelengths.

This being so purpose-built vessels incoporating the skimming function in as a rigid integral element in their construction will be inoperative in waves. This conclusion means that great cost savings are immediately possible, since such vessels are extremely expensive to build end maintain and would if chosen be required in large numbtrs in order to overcome the enormous logistic difficulties inherent in this concept.

Another system which can be reported on favourably is that developed by ORI/Star Offshore, known as the Force 7.4. This is a single-ship system and is based on the flexible absorption rope principle. Consequently it has excellent wavefollowing characteristics and should recover oil of viscosities allowing adhension to the rope at rates controlled only by encounter rate.

These two systems meet the identified requirements is different ways. They may differt, however, in encounter rate potential, the Sprirgsweep system having a collection boom while the Force 7 does not. On the other hand the Force 7 may be capable of operating at a greater ship speed.

It is believed that further evaluation of these systems can now be achieved only through operation at real spills. Decisions can then be taken on a cost -effectiveness footing in the light or considerations present to here.

The Eleni. V. incident involving a spill of heavy fuel oil which was solid at sea temperature, represented a particularly difficult case. The Antonia Gramsci incident, however, revealed that solid oil in the presence of ice presents even greater problems.

In the end some 400 m of pure oil was recovered at sea (the actual recovery rate is nc t, available) and some 400-500 m3 was subsequently recoved from 3C0 km of coastline at a cost of 17,500 m^3.

Results of Investigations on Remote Sensing Techniques for Oil at Sea

It has been shown that oil slicks on water are detectable

by observing reflected light in the near-U.V. visible and nearIR regions of the electromagnetic spectrum.

In addition, slicks may be observed in the thermal IR region because of apparent differences in surface temperature as between oil and water.

In order to select the best wavelength windows for operation, however, it is necessary to codsider the problem in greater detail. Apart from the surface-reflected raidance in which pne is interest, There are two other background *components* contributing to the total radiance detected by a remote sensor.

These are a volume-reflected component and a path radiance scattered by the intervening atmosphere into the sensor. This latter components depends on path length at higher altitudes can be the predominant effect.

In addition, the sources of the reflected radiance have to be considered, viz, the sky radiance which is reflected at the surface component and the so-called global radiance which is the sum of the sky and sum radiances and is received by the sensor as the volume and atmospheric path radiance.

Thermal radiance devies measute apparent temperature which arises from the physical or 'real' temperature and the thermal emissivity. The presence of an oil slick can cause changes in both parameters. In the case of physical temperature an oil slick could in principle produce the following effects

- The layer could absorb solar radiation and become hotter than the underlying and surrounding sea.
- Oil close to spill source could retain heat and thus be better than the sea.
- Evaporation of volatile components could have a cooling effect on the oil relative to the sea.
- Oil could reduce evaporation of water and thus produce a warming effect relative to the sea surface where water /would continue to evaporate.
- Oil could restrict aeat transfer between the

atmosphere and the sea so that in warm weather the sea would be cooler under the oil and in cold weather it could be warmer than it would otherwise be.

— The movement of wind over the oil slick could warm it up by a mechanism of viscous dissipation of energy relative to the sea surface.

From the above list of possibilities it is clear that oil slicks at sea could be hotter than sea water or cooler depending on the final outcome of a variety of processes individually giving rise to conflicting results, water.

Oil is variously reported to have an emissivity 0 02-0.005 less than water, which may be expected to give a temperature contrast of 1-3 degree K.

Low Light Level Television

This equipment is well worth investigating as a remote-sensing tool for oil on water. If satisfactory performance can be demonstrated one immediately has a number of advantages, via low cost and ease of operation together with real-time display of the imagery, LLLTV can have a spectral response in the 250-1200 nm region of the spectrum depending on the characteristics of the tube used. Oil has a greater surface reflectance than water in this region and so it should be possible to see oil in contrast to the water.

Infrared

Some tank and wind tunnel studies have been carried out to investigate possible mechanisms to explain the difference in physical temperature between oil slicks and the surrounding and underlying water 5. In these experiments, thermocouples were places in the oil and in the water immediately below the surface at 1 mm depth and in the bulk water at a few centimetres depth. Control tanks containing only water were similarly equipped with thermocouples at t mm and at a few centimetres depth.

Line Scanners

Ultraviolet and infrared line scanners, have received

considerable attention in a number of countries and the Daedalus system features prominently in this work.

Canadian and US workers have evaluated the Daedalus 1230 dual-channel line scanner which is fitted with a UV detector sensitive in the 270-370 nm region and a thermal IR detector for the range 850-12.50 nm and the Daedalus 1260 multi-spectral scanner is which the near-UV to near-IR region, 380-1100 nm, is divided into ten channels and there is in addition the usual thermal IR part at 850 -1250 nm.

Ultraviolet

The UV imagery from the dual-channel unit gave good contrast for the thin regions of the slick. This device had a better performance than the multi-spectral scanner mainly because of the lower noise equivalent radiance value for the UV detector.

This fact indicates that features other those such as detector sensitivity and noise levels will affect results and to some extent therefore the results obtained apply only to the system used.

Microwaves

Microwaves sensors are available with wavelengths in the range 0.1-100 cm and can operate either in the passive mode, measuring difference in brightness temperature between oil and water at a given wavelength, or in the active mode as radars receiving a back scatter from the sea or oil surface.

Radar

Waves at sea are of two types, long wavelength large amplitude gravity waves and small capillary waves, the latter being the first to be produced as wind speed increases from zero and which ultimately are super imposed on the larger gravity waves.

Radar backscatter is mainly due to these capillary waves, but imaging radars can see gravity waves because of the increased concentrations of capillary waves on the downwind faces of gravity waves as compared to the upwind sides.

This produces a periodicity in the radar backscatter of capillary waves even though the radar wavelength is very much smaller than that of the gravity waves. This is turn suggests that backscatter intensity will vary with viewing direction with respect to wind direction.

Microwave Radiometry

Microwave radiometry is a passive technique which selects changes in microwave brightness of the sea in the presence of oil slicks in a manner amalogous to the infrared technique. It also offers the possibility of measuring oil slick thickness with the added advantage over infrared that it is independent of atmospheric water content *i.e.* fog or rain.

Calculations of microwave brightness of oil over water for a given frequency predict alternating maxima and minima as the oil thickness increases. By using two frequencies, ambiguities can be removed and thickness uniquely determined over a wipe range. Measured antenna temperature is the average of the brightness temperature over all directions weighted by the antenna response pattern.

The total brightness temperature in any direction is due to the radiation emitted by the surface, the downward radiation of the sky reflected by the surface, and the emission and attenuation of the atmosphere between the surface and the radiometer.

The quantification of oil in effluent waters will be discussed. In both these situations the source of the oil is known and it is the quantity of hydrocarbons in the water which is being sought. In the aspect now under discussion it is the source of pollution which is to be established.

This requires a different approach. Before discussing these different requirements, however, it may be worthwhile to recall who analysis may be necessary in this context. Aerial surveillance with standard photographic equipment is not in itself sufficient to establish that a named ship identified in a photograph, is actually responsible for discharging the oil which appears to emanate from that ship **in** the photographic evidence.

In fact the best way to establish that the vessel is discharging oil is to sample the area which shows as an oil slick in the photographs and to characterise that sample by chemical means in such a way as to show that it is identical to a sample of the oil carried on board.

It is worth recalling also that this comparison between pollutant and putative source may be advisable even in the case of a shipping casualty where attribution of oil source might be thought obvious.

This can rise when passing vessels take advantage of the widespread pollution arising from casualty in a given area to discharge tank washings, bilges, etc. As a result some of the oil arriving on shore may be erroneously attributed to the casualty and the operational discharge offenders may escape detection.

Sampling of the Oil Cerried on Board Ship

It has alreaey been stated that the best way of establishing that a pollution sample orginted from a suspect ship is to compare the results of chemical analysis of the pollutant with that of the oil carried by the ship. In fact it is the only way.

Such is the complexcity of individual oils and the wide variety of potential sources of distinct oils that it is quite impossible the basis of a pollutant sample to state the source of that pollutant. Generally speaking, the best that can be done with an unreferenced sample is to say that it is a fresh crude oil, a weathered crude oil, a light or heavy fuel oil or tank washings.

This quite insufficient to associate it with a particular source. This being so, it is necessary in a case of suspected operational discharge to sample all the oils on board including bunker fuel and lubricating oil. This may involve sampling all the tanks, though reference to the cargo manifest may help to reduce the amount of required sampling. Similarly in a shipping casualty it is useful to sample all oils on board since damage of the vessel can result in the release of oils.

Sampling from the Sea and Beaches

In collecting samples of liquid oil from the surface of the sea, scare is necessary to ensure that oil is indeed being collected and that the result is not samply of dirty water in an oil contaminated container.

Where the pollutant is a relatively thick layer of water-in-oil emulsion it is easier than if the pollutant is fresh oil present to about 0.1 mm thick or even less, but in all cases it is easier to collect water than oil. The use of a container with a bottom outlet and tap is helpful in collecting sufficient oil since with such a device water can be drained off and sampling repeated if necessary.

A funnel can be used in this way, using the finger to relase the water and retained the oil, Alternatively where the oil layer is very thin or is in the form of many small isolated globules of oil or emulsion, it may be possible to use adsorption pads, These may be of Teflon, steet or polyurethane.

The oil sticks to the pad when it is placed lightly on the oil-contaminated sea surface. The pad is then placed in a sealed airtight contniner transport to the laboratory. Oil can be removed from the pad by use of a suitable solvent and this operation can be carried out either before transportation to the laboratory or after arrival there.

The Nature of the Oil Spill Response Problem in Inshore Waters and Beaches

Everything that has been said in the chapters n at-sea oil spill response holds also for inshore waters. However, there are a number of additional points of specific relevance to the inshore situation.

As oil approaches the shore, the efforts made to combat it either by dispersants or by mechanical recovery means will of necessity be conducted closer and closer to shore, and if these efforts are inadequate the oil will eventually strand on the shoreline.

The extent to which the oil strads will depend on the extent to which it is treatable at sea and this in trun will depend on soil size, oil type and the proximity to shore of the original spillage.

So far, beach-cleaning costs per tonne are con%idred to be so high as to justify operations at sea. For the moment it will be assumed that if activities at were justified, they presumably still are close to shore.

The next step is to consider dispersants and mechanical recovery separately with a view to establishing whether their ought to be an reasons for discontinuing activities commenced at sea simply because the oil is coming closer and closer to shore, and alternatively the extent to which it might actually be easier to carry out the desired operations now that the oil slicks way be reaching more sheltered waters.

OIL RECOVERY

In cases of oils which are not amenable to dispersants, only recovery is possible. As with dispersants, the establishment of water-in-oil emulsions with their increased viscosity will rend r the recovery of oil more difficult.

On tha other hand, the possibility of thicker layers brought about by the booming effect of the beach itself (the best boom available) may provide greater I syar thicknesses and therefore the possibility of increased encounter rates for oil recovery systems. In addition, unless the inshore waters are exposed to the full force of the weather the general sheltered nature of many inshore situations will assist in oil recovery operations.

In general, therefore, provided there is enough sea room and depth of water for ship mznoeures, the deep-sea equipment will be usable in sheltered inshore conditions.

However, in addition, because of the shallower water depths it may be convenient to more booms rather than to low them. Moore of booms, being static of course, do not increase encounter rate but are used to protect specified stretches of shoreline and/or to deflect moving oil to a selected recovery point.

These techniques are specific to the inshore situation. In general, mechanical recovery of oil and water in oil emulsions should be easier in sheltered waters than at sea. Indeed it may be better to view oil of recovery as a sheltered water technique which may in certain circumst-ances by usable at sea rather than to present it as a sea going technique which can automatically be used in sheltered environments. Be that as it may, boom utilisation and oil recovery techniques in relation to the particular inshore situation.

Where sea room and depth of water, limits such operations there is scope for smaller devices deployable from smaller vessel, e.g., the net system from inshore fishing boats. However, the use of small vessels is not without its difficulties.

The logistic problem arising from lack of storage on board mitigates against the small vessel approach. Again, if the oil is already against a harbour wall or other coastal structure it may be simpler to deploy the oil recovery equipment directly from the land.

There is little point in working from a small boat if one can work directly from the river bank or quay wall or shore. Recovered fluid is then already ashore and one step in the logistics of waterborne operations is immediately eliminated.

In considering how best to deal with oil once it has come ashore it is necessary to have regard to the type of beach material affected. Beach material is classified in terms of grain size.

Mud is The term used to describe grain sizes less than 0.05 mm. Mad beaches generally have a slope of less than 1 degree they develop in areas of low water velocity which permits the deposition and retention of such fine material.

Such beaches are saturated with water at all times and becaus f the low drainage which this implies, the surface is usual covered with a thin layer of water. They have very low 1 d-bearing capacity and are generally speaking of high biologic I value.

Sand beaches have a particle size range from 0.05 to 2.0 mm with bench slopes between 1 degree and 40 degree. Such beaches are generally more exposed than mud beaches and are subject to seasonal erosion and deposition cycles depending on the energy of the incoming waves. Drainage rates are higher than for mud out again permanent saturation and surface water layers may be encountered.

Load-bearing capacity is higher than for muds, sand beaches are of low biological value. Gravel and people beaches are of particle Size 2.0-50 mm and 50-256 mm respectively and tend to be narrower and steeper as the particle size increases from gravel to pebble. Such beaches are again of low biological value.

Apart from these physical difference due to particle size, other parameters are relevant such as exposure of the beach in question, the onshore energy levels of the waves to which it is exposed and the accessibility of the beach from the point of deploying pollution response equipment.

CRITERIA FOR THE SELECTION OF MOORABLE BOOMS

It is important to distinguish the specific uses of booms. In the first place the booms may be used in a sweeping mode to concentrate oil towards a skimmer so as to increase the oil encounter rate of that skimmer. In the second place booms may be used to protect certain coastal resources from oil contamination or to deflect oil on so a nominated or sacrificial beach for recovery.

In the former case the booms are deployed from ships and are used in a mobile mode. In the latter case they may be, therefore used in a stationary mode. Of course many aspects of boom design are common to both applications, but it is as well to be aware of those aspects in which boom design in specific to one situation or the other. Some boom are usable in both the towards and moored conditions but others are not.

It will rarely be possible to deploy moored booms in such a way as to protect the very large lengths of open constline which may be threatened by oil pollution in a major incident. The use of such booms must be selective.

In addition, depending on time-scales, logistic problems, etc., it may be impossible to have such booms in place in the required time.

However, in some cases it will be possible to consider these matters as part of an overall plan for cargo transfer in a safe haven. In such a case the deplopment of any necessary booms will require to be completed in good time for the arrival of the casualty.

The selection of towed booms for use at sea has already been discused in other section of this chapter. It is now time to discuss criteria for the selection of morable booms for coastal protection.

The objective is to reduce the range of possible boom designs which need to be considered and to identify those elements of boom design which need to be convested in a boom specification.

Boom Specification

If a stationary floating barrier is placed at right-angles to flowing water, the water must of necessity pass beneath the barrier.. It is found that water passing in this way beneath the barrier.

It is found that water passing in this way beneath floating booms carries oil with it if the flow rate normal to the boom is greater than 1 knot.

It seems unlikely that performance can be raised significantly be attention to detailed boom design since the limit appears to be quite fundamental. What is certain, however, is that it is quite possible to design a boom the performance of which will fall short of the limit for quite avoidable reasons.

Boom design therefore is a matter of ensureing that the final product will.

- — Attain the fundamental performance limit
- — Maintain this performance in waves and carrents when moored in position.
- — By of adequate strength to withstand the loading forces involved.
- — Be light enough to handle in deployment, use and recovery.
- — Be of low enough bulk to facilitat storage and transportation.

Oil Recovery in Inshore Waters

1. Encounter rate, 2. Reduced performance as viscosity increases, and 3. Reduced performance in waves. If these three problems cannot be solved to a satisfactory extent, some oil will come ashore.

The problems of oil recovery in shelteres waters and close to shore, in docks, harbours, etc., are rather different. To begin with, the urgency to remove the oil before it reaches shore is now gone. It although obiously the rate of removal relates to cost per tonne of oil removed and to that extent remains important.

As it happens, however, the oil pressed against the shore or other beach structure may be in a much thicker layer and therefore allow very satisfactory recovery rates. Secondly, reduced performance because of increase in viscosity may not be so significant since, as stated above, the urgency may have gone but of course costs remain important,

Thirdly, waves will be much reduced and in certain circumstances totally absent. However, whereas at sea ene is content to sweep the thickest layers using booms and to leave the thinner regions, and while in sheltered waters booms may still be used to reduce the area over which the oil is spread so as to maintain an adequate thickness or recovery, there will be situations in which it will be essential to be able to collect oil which is present in only very thin layers and eventually to effect complete removal of oil from the water surfaces.

SKIMMERS FOR GENERAL INSHORE USE

Weir Skimmer (Slurp Skimmer)

Perhaps the most basic oil skimmer of all is that developed by Esso Research centre and known as the Self-Levelling Unit for Removing Pollution, of SLURP 2. The SLURP, is a weir skimmer with no moving parts.

It is of small size, 935 mm in width and weighing 28 kg, and can operate in water depths as little as 25 cm. The attitude of the floating seimmer and consequently the depth of the weir depend on the amount of liquid contained in the unit.

Since high pump rates are appropriate for deep oil layers and low pump rates for shallow layers, the device is designed so that a high pump rate maintains a low loading and therefore a floating attitude which results in a deep weir position.

Conversely a slow pump rate maintains a high fluid loading and consequently a floating attitude which provides a shallow weir position. Clearly, to achieve the best results the skimmer should be used with a variable-speed pump.

The Komara Miniskimmer

One of the commonest manifestations of the principle of preferential adsorption is that of the adsorption disc which is the basis of the Komara Miniskimmer 3. This device is well worth a study in its own right but, being typical of a number of such devices employing disc, the results obtained can be of general application.

Disposal of Recovered Oil

Oil may be collected from water surfaces or from a variety of beaches comprising different types of solid material. So far in considering the problems of oil spill response we have considered the properties of spilled oil in relation to the means of response, viz, recovery or dispersants.

It is now necessary, assuming that oil has been recovered, to consider its properties and levels and types of contamination from the point of view of tempory storage, transportation and final recycling or disposal.

Oil collected from water surfaces, even if originally of low viscosity, will have lost its more volatile fractions by evaporation and will almost certainly have formed water-in-oil emulsions by the time it is collected. As such it will be extremely viscous and, being non-Newtonian or thixotropic in properties, it will be virtually solid when allowed to stand in storage of any kind.

The main contaminant is, however, water, so that if the emulsion can be broken there is a good chance that the oil fraction could be put to some use. Indeed if the recovered emulsion is to be handled through storage, transport, etc., it is essential that the emulsion broken anyway.

Oil collected from the beach is another matter entirely. It may contain sand, single, pebbles, stones, sea wood and the whole range of potential beach debris either separately or together. Indeed because of the difficulty with graders and scrapers in skimming the desired thin layer, material collected from beaches tens to be oil contaminated in this form is to be reprocessed or otherwise put to useful purposes, it entire mass must be disposed of by some means or other.

If it is intended to reprocess recovered oils through a refinery it is important that they are not contaminated with dispersant. The identification by analytical means of dispersants in oil is difficult but under investigation at present. The best means of avoiding the problem, however, is to ensure that when the emulsion becomes established to the point where dispersants are no longer effective, their use should be immediately discontinued.

If they did, the dispersant would be effective. When they do not simply wash off into the sea and become diluted to harmless levels. There is some talk at present of intentionally gelling oils to render them mo-e easily collectable in nets, i.e., without extrusion, if their properties are already such that other means of collection are precluded by high -viscosity Just how it is proposed to mix the gelling agent into such highly viscous materials under the conditions prevailing at sea is as yet quite unclear.

However, what can be stated clearly is that such agents ought not to be of such a type as to give rise to problems with either final disposal or reprocessing.

Demulsification

It has been stated elsewhere that crude oils which spread on the sea surface after spillage furm water-in-oil emulsions in which oil is the continuous phase and the water is present as small droplets dipersed throughout the oil.

Such emulsions are more viscous than the ·riginal oils and have non-Newtonian properties, i.e., they are thixotropic and their flow rates are a function of the stress applied.

Thus under the influence of wave action they may have viscosity which permits pumping, but once In a storage tank they will increase in viscosity by standing still and be virtually impossible to remove from the tank when the time comes to do so.

General Problems of Oil-water Separation

A dispersion of two immiscible liquids is thermodynamically unstable tenuing to reduce its free energy by reducing the interfacial area between the two phases by droplet coalescence.

Droplet coalescence occurs to limited extent between rising droplets, but eventually coalescence of droplets in the more densely packed region under the phase boundary and with in the growing continuous phase itself assumes more importance.

Finally two separate coctinuous phases are formed, the separation having been achieved by gravity acting on components of differing density. Emulsions by contrast are quasistable systems in which coalescence is inhibited, often by the presence of a third component, present usually at comparatively low concentration and which is adsorbed on the droplet surfaces.

This component is known as an emulsifying agent or emulsifier. Naturally occurring emulsifiers give rise to water-in-oil emulsions.

Although dipersions tend to separate by gravity, they will not readily do so if the droplet size is so small that movement towards the continuous phase boundary is interfered with by Brownian movement within the continuous phase in which the droplet is located.

The term 'secondary dispersion' is used to describe very small droplet size dispersions. These are responsible for persistent hazes in a number of industrial processes and require special techniques for separation.

Chapter 8

CONTROLLING MARINE POLLUTION

There are many factors, acting both individually and in combination, which govern the effect of oil on marine life. These will include the type and quantity of oil, the direction and quantity of pollution experienced by the organism, the state of the oil, a g., fresh, weathered etc., the season, with respect to the annual cycle of the organism, and the habitat and the natural stresses to which the organism is subjected.

Any or all of these can affect the organism in many different ways. The effect may be lethal or sub-lethal, in the latter it can disupt physiological or behavioural activities and death may result due to interference with feeding of reproduction. The uptake of oil can result in tainting or possible carcinogenesis either of the organism contaminated or one higher up the food chain.

The visible and well-known effects are beach pollution and the death, due to oiling, of seabirds. It is not unreasonable to wonder, if not actually to fear, whether oil might nor adversely affect marine life and, in particular, harm fish the source of much food for min and his domestic animals either by reducing their numbers of rendering them unfit to ear.

Like all life, life in the sea depends upon solar radiation for **its** source of energy. Through photosynthesis, marine plants, of which by far the most important are the minute drifting organisms collectively know as phytoplankton, store this energy in a form which other organisms such as zooplankton and the fish which prey upon it can utilies, a simplified diagram showing the board outling of what is in reality a highly complex food web. It is helpful to realise that at each step the efficiency of conversion of food into body weight is only about 10%. Nelson Smith estimates that man would require at least on kilogram of phytopla-nkton to gain one gram in weight.

This is a useful statistic for it highlights the fact that direct effects on fish stocks eaten by man are of much greater economic importance than equivalent damage in terms of biomass lost, to the planktonic organisms on which the fish depend.

The food web diagram also highlights the fact that fish, and other organisms well removed from the primary production stage represented by the phytoplankton, have by proxity, consumed relatively enormous quantities of organic materials some of which, although only present in small concentrations in phytoplankton, may collect at the higher tropic levels of the foood web.

The question of whether there is an accumulation of petroleum hydrocarbons compounds, if they represent a credible threat to the resource or the consumer, and the other possible effects on marine life of oil and of chemicals and methods used to clean it up.

PLANKTON

At almost all major oil spills, lay men and ecologist alike express fears about possible harmful effects on plankton. Concern can usually be summarised under three headings

— A layear of oil on the surface will, by* occludiug light, cause a serious drop in photosynthesis by phytoplankton.

— The oil layer will interfere with gaseout interchange at the sea surface and dissolved oxygen levels will thereby be lowered.

— Oil and dispersants will exert serious toxic effects on plankton.

How real are these effects and what is their ecological significance ? The first and second are the indirect consequences of meansurable physico/chemical phenomena and will be treated first.

Occulsion of Light

Ottway studied the light transmitting properties of 20 crude oils made up as 5% solutions in white sprit and shoed that different crude oils vary widely in this respect. Oils high in asphaltenes and sulphur were found to absorb most light and, for all oils, the absorption spectra exhibited maximum at the blue end of the spectrum (325-350 nm), indicating that the 44 same group of compounds is mainly responsible in all of the oils tested. A number of photosynthetic pigments found in phytoplankton also have absorption maxima at the blue end of the spectrum.

Potentially, at least, their photosynthetic activity could be impeded by crude oils spilt at sea. It is important to remember the scale of the problem. A huge slick of black oil covering 2000 square miles of the North Sea actually covers less than 1% of its area. It is safe to conclude that occlusion of light by spilled oil, a phenomenon easily demonstrated in the laboratory, has no ecological significance for the plankton of the open area.

Effects on Dissolved Oxygen Levels

The solubility of oxygen in most hydrocarbon mixtures is similar to that in sea water. An oil slick is not, therefore, an impenetrable barrier to gaseous interchange. For a full discussion of this issue, the reader should consult Nelson-Smith; suffice it to say that in a 2 day trial, Brown and Reid found that boild water, beneath a 17 mm layer of oil, absorbed 73% as much oxygen as the control.

Confirmation that oil slicks do not interfere significantly with gaseous interchange at the surface of the open sea is provided, by an observation by e Scripps Institution of Occeanography, on dissolved oxyg levels below a heavy oil slick in the Santa Barbara Channel. The oxygen saturation beneath the slick was 98.5% of that i water outside the slick area.

Toxic Effects

There is abundance evidence that freshly-spilled ci ude oils contain low boiling point substances which are actually toxic to the planktonic organisms. The most toxic of these are water soluble aromatic derivatives. Low molecular weight alkanes (10 carbon atoms or less) may induce narcosis in high concentrations but these are unlikely to result from a crude oil spill in the open sea.

The aromatic solvent, used in early emulsifiers were also found to be highly toxic to planktonic organisms at levels of less than one part per million. In letter formulations of equivalent dispersing power, the use of non-aromatic solvents has reduced toxicity by over two orders of magnitude.

Useful as laboratory assessments of toxicity can be for comparing the acute toxicity of oils and dispersant chemicals to small number of organisms they are rarely a refiable guide to effects on wild populations. In that such tests are necessarily short term they under estimate toxicity, but on the other hand, the toxicant concentrations used in laboratory experiments are rarely maintained for any length of time in the sea. Evoporate losses the rapid dilution of residual, soluble compounds ensures than toxic effects are short-lasting, even for high toxicity, solvent emulsion of the aromatic type.

Thus although highly localised short-term effects on plankton were observed following the massive use of high concentrations to disperse Torrey canyon oil. There was no evident that any planktonic speies was significantly affected the population level.

The Torrey canyon experiance was quiete exceptional in the amount and toxicity of the materials used to dispense the oil, and even these it would seem that the rapid evaporation and dilution of the toxicity combined with the dispensed nature and rapidturnover of the plankton to minimise the ecological impact.

Por other major spillages, including the notorious Santa Barbara blow-out the toxic effects on the planktonic tissuse have proved to be virtually detectable. The plankton of milfod Haven was examined to see if the development of a major oil port had any effect. The report stated that changes which have occured are no greater than might have been produced by natural environment changes and population dynamics.

Sanbourn suggests however that the effect on benthic larvae when they are in upper layer of the ocean may be more longer lasting as benethic animals required years to reach maturity and so the effect on the larvae may endour longer. The only planktonic species which may conceively be in any ecologically significant danger from the toxicants associated with crude oil or dispersants are those which form fairly localised aggregations.

Arrow worms fish eggs, and larvae may come in to this category. The likelyhood of ecologically significant effects in such cases is examined in detail. Finally, to keep our assessment of toxic effect on plankton in their true perspective it is helpful to consider the scale of plankton production of in the sea relative to the that of petroleum spilled.

Fraser estimates annual world phytoplankton production to be about 150,000 million tons and that of zooplankton 15,000 million tons. By contrast estimates of annual losses of petroleum in the vary from 2.5 million to an improbable maximum of 10 million tons. Only a fraction of this petroleum is toxic to plankton and very little of this is ever likely to come into contac: with planktonic organisms.

Fisheries

As was said earlier, the purpose of this chapter is to arrive at an asse»m.nt of the effects of oil pollution, an environmental stress, for which man is largely responsible, on the biology of the sea. It is important, in this context, to consider the particular case of sea fisheries, a sector of the marine enviroment in which man already exerts a direct and measurable influence.

The term, 'fisheries', has a variety of meanings, the widest embracing all exploited stocks of marine organisms from macro-algae to whales, and the narrowest exploited species of fish alone. In this chapter, the term is taken to mean the fish and shelfish of commerce.

Oil pollution is potential treat to these resources for three main resaons

— Pollution any directly harm fish and shellfish stocks.
— Pollutants adhering to, or accumlated by, fish and shellfish may render them unacceptable to the consumer.
— The presence of spilled oil may interfere physically with fishing operation.

Controlling Pollution

Now that the nature of pollution and s-)me of the sources of this undesirable product of society have been considered, let us concern oursveles with some of the avenues open to us to accomplish some sort of control.

The most obvious control strategy that can be used is to some how or other completely prevent the introduction of the pollutant into the environment. This can be done by not producing the pollutant or by restructuring the process so that the pollutant is no longer a byproduct.

Let us consider a few major sources of pollutants and some possible methods of control in these particular cases. The most obvious solution to the problem of maintenance redgi is to cut off the sediment supply to the areas where se(iimentation is a major problem. This can be done in a number of different ways.

The most effective of all is to prevent upstream erosion, but at the same time this is perhaps the most difficult. The entire watershed must be policed very carefully to make sure that the natural landscape is change to such an extent that erosion no longer occurs. This means extensive terracing and replanting of areas that have been denuded for on reason or another.

Of course, this also means that extreme measures must be taken in areas that have no natural cover, such as agricultural areas or areas where contruction is underway. A certain amount of this is being done and these are a number of legal requirements for persons engaged in construction or agricultural activities.

Another method of preventing sediment from reacing the undesired area is to provide settling ponds. If a stream is identified that carries a large sediment load, a pohd and dam arrangement can be built so that the velocity of the stream is decreased to zero and the suspended sediment will have a chance to settle.

The water is released after it has dropped the major portion of its load and is allowed to proceed downstream The solution since the ponds will fill up rather rapidly, requiring either building of new ponds or dredging of some sort.

Sometimes the sedimentation that occurs is channel area is a result of sediment loads that are introduced into the marine environment from nearby rather than those carried from far upstream.

This occurs, for example, when a channel is dredged with unstable sides so that the material just tends to slough into the deeper regions.

This requires continual dredging as removed sediment is almost immediately replaced by bottom material from the sides. One method of correcting this situation is by stablization of the bottom. Adding larger, more dense material such as gravel to the bottom is a technique utilized occasionally, although this is usually more expensive than dredging so that it is not done too often.

Bottom stabilization, however, is a technique that is being used more and more to prevent the movement of bottom materials within relatively small contiguous areas.

Oil Pollution

Another problem that, to a certain extent, appears amenable to prevention at the source is that involving oil. Unfortunately, the sources of oil in the marine environment are extremely difficult to delicate. Many figures have been published and, in almost all cases, there is a significant disagreement as to the amount of oil discharge into the marine environment and the sources of the oil.

There is no doubt, though, that a large portion of this amount originates on land as waste product. As would be expected, the major source of this is associated with the automobile. Some of this lost oil is dropped from leaks in cars and trucks, some is deliberately dumped as sed crankcase oil, while some results from the deposition of,unbu ned exhaust gases.

From roads and streets this material is was marine environment, usually without benefit of treatment. Even when passed through treatment plants, however, petroleumrich effluents still provide problems as they are difficult to treat and they tend to make treatment for other pollutants less effective.

All the experts agree that the greatest terrestial source of petroleum products in the occean is used crankcase oil. Some go so far as to indicate that this source supplices more petroleum to the sea than any other source. This may be true because within the last decade or so the habits of American motorist have change markedly to the extent that a large portion of them now change their own oil, discarding the used oil in a manner that probably causes it to end up in storm sewers.

The major reason for this probably is the almost complete disappearance of the used lubricating oil re-refining industry in recent years since it was not economically feasible to refined oil. However, with the modern continual increase in the price of new oil, the re-refining industry is having a rebirth so that a large portion of the oil previously thrown away will now be reused.

Another entirely different source of oil in the marine environment is that resulting from accidents that occur during shipping and transferring processes generally speaking. The more handling steps involved, the more accidents and associated losses can be expected.

The figures as to how much is actually lost in handling accidents compared to thas lost in normal ship procedures, such as pumping bilges, is not well known, but there is not doubt that a very large portion of the oil reaching the marine environment is the result of poor handling procedures in both the operation of vessel and the transfer of oil from one carrier to another.

Better procedures are need, but perhaps even more than better procedures, structure enforcement of accepted safe procedures would go a long way to decreasing the amount of oil that enters the ocean each year.

When a tanker emplies its load of oil, the dynamic characteristics of this vessel are markedly changed. Its weight is decreased about a thousand times and therefore it becomes extremely difficult to drive the ship, since it was designed to be driven with a full load. Therefore when most tankers do not have a load of oil, they will fill their tanks with water for ballast.

When they return for a fresh carg of oil, they must pump this water from the tanks, releasing large amounts of oil as the residue left in the tanks is washed out. There have been many schemes suggested for decreasing the amount of oil that is wasted in the ballasting-unballasting process ; however, it is not clear at this time how effective these processes are or how effectively any requirements can be enforced.

Modern supertankers are extremely large vessels and therefore have major structural weaknesses so that occassionally they will break up in unusually heavy seas. They are simply not designed to be utilized under unusual stom conditions, so that there will continue to be losses of this type in rough weather.

In passing, it should be mentioned that once oil is spilled at sea under large wave conditions there is no way for this oil to be retrieved. On the other hand, oil may be retrieved inharbor areas where the sea surface is relatively calm, but only when there are waves no higher than approximately one meter.

The obvious way the only way to prevent oil from getting to the marine environment when a ship founders is to prevent the accident from occurring in the first place. Alternatively, once the accident has occurred the oil must be prevented from escaping from the ship. It is somewhat doubtful whether these solutions will ever be possible.

Recycling and Reclamation

If it is impossible for one reasons or another to prevent the introduction of a pollutant into the system, it may be possible at some place along the line to extract this pollutant and resue it. It should be kept in mind that although "consumer" is a commonly used word, there is no such thing as a consumer.

The word tends to imply that material is being used up, while, in actuality, it is just borrowed for a while and returned. It may be in a somewhat different form or perhaps a somewhat different shape, but nevertheless it is still returned.

We don't destroy any of this mass; we just change its farm, so that the general concept of recycling or reclamation is extremely appealing especially when we consider our larger solid forms of waste. If we could find some way to utilize al of these materials, we would solve a number of our pro lems especially those involved with space.

Sometimes, though, it is not necessarily desirable to solve a space problem sometimes the space problem works to our advantage. For example, it has been found that biological communities living in the sea require a certain amount of physical shelter.

Some species just like to be along, whereas in other cases the young of the species require places to hide to protect against predation. A marsh or wetlands area is a good breeding place because both of these conditions prevail. It has been found that in many cases these conditions can be created artificially by introducing large juck articles such as old automobiles or worm out ships to form effective artificial fisning reefs.

There are disadvantages to using waste products for this sort of activity. One is that in many cases these junk objects are transported to undesirable areas by dynamic oceanic, forces. Pollution is not controlled when automobile fenders and doors wash up on swimming beaches. When creating a fishing reef, steps must be taken to retain the integrity of these reefs for a long period of time.

If proper attention is paid to the magnitude and nature of oceanic forces, artificial reefs can be made relatively permanent. In actuality the whole recycling and reclamation business is predicated on cost. If it is economically feasible to recycle or reclaim meterial, it will be recycled or reclaimed.

If it costs more money to use the recycled material than new material. than the material will not be recycled. It is difficult to imagine that certain materials, such as glass, will ever be recycled to any great extent because silicon is one of the most common elements in the earth's crust and it costs just about as much to make a new bottle as it does to recycle an old one.

However, in the case of aluminium there is about a factor of five inv Aved in recycling coast as opposed to manufacturing from scratch. With the present state of the economy where things become more expensive from year to year, it appears that more items will find themselves on the list of materials to be recycled or reclaimed.

Industrial processes will change as it becomes more feasible to extrat materials for the waste stream from a cost standpoint and, at the same time, industrial waste streams will become less and less polluted. Even in the area of multiple waste product use, such as the combination of water heat from either manufacturing or electrical power generation and waste nutrients from sewage treatment plants for the purposes of aquaculture, cost always be the controlling factor.

It has been known for some time that evaludted temperatures and the artificial addition of nutrients will increase growth rates of many organisms markedly. The choice of which organism to grow is involved not only with which would be best to grow, but also which there is a market for.

Setting up an aquaculture system involves picking organisms which respond favourably to the environment, developing a system by which the optimum growth rate can be produced, developing a marketing system so that the end product will be sold at the maximum possible price, and assuring the continuing existence of market for everything produced. Not only is science included but also sociology and economics mut be considered, and it is probably in these latter two areas that the major difficulties reside.

Consequently, no sophisticated aquaculture systems using both waste heat and nutrients have been developed at this point, but it is probably only a matter of time before we will have extensive aquaculture using man's waste products.

STORAGE OF POLLUTANTS

Another method of controlling pollution, utilized for certain pollutants, is to store them at some distance from man's activities so that they are as inaccessible as possible to the major fraction of the population and pathways to man. Thus we essentially throw them away, but we thrown them away, in a place where no one can get to them and their effects cannot get to man.

Occasionally nuclear waste products have been dumped in deep oceanic areas in the hope that when they do break out of their canisters, they will not find can easy pathway back to man since the de er life forms are probaly not in any food chain involving an. Perhaps not so well known is the fact that deep ocean disposal has been used for other materials such as average ordiance and poison gases for which the military has not further use.

In the case of nuclear products, there is a very vociferous school of thought that believes storage facilities should allow waste retrieval if a technological bre: k through is even accomplished that would allow these nuclear waste products to be profitably utilized. Another school feels that they should be placed in an areas as completely inaccessible as possible to everyone.

This latter school in tuitively seems the safest, but, on the other hand, most common storage places are just not the type. There is always a small probability of a pathway back in man being found. One suggestion for a nonretrivable storage place which holds some promise has been made by Issac Asimov.

According to moder Geological theory the earth's crust is composed of a number of separate plates, all moving with respect to one another. In some places the plates are moving together and in others aprts, following the motion of the mantle, the layer of material underneath the earth's crust. As a result of convection, the mantle tends to rise toward the earth's crust in certain areas, producing oceanic ridges such as the North Atlantic ridge.

At these divergence areas, where the mantle material is brought up to the earth's crust, the motion becomes horizontal and the crustal plates are carried away from the ridges at a rate of a few centimeters per year. The convergent regions where the plates come together are where the mantle material is carried back down toward the centre of the earth.

In direct contrast to the regions of divergence associated with ridges. the convergent areas are associated with oceanic trenches where crustal material is carrie3 down and under the continental plates. These oceanic trench areas the very deep, often as decp as 10,000 meters, and at the same time are actively engaged in transporting material into the bowels of the earth.

Controlling Pollution by Zonec

Another method of controlling pollution is by specifying certain areas that can be utilized for effluent discharge and other areas that cannot. This method has been used on hand as records have been kept. Although zoning of certain regions as industrial, commercial, or residential is a relatively recent innovation, there always was city dump.

Now-a-days we are even starting to convert our city dumps into usuable areas, such as gulf courses or sky slopes by convering, planting and landscaping as rapidly as possible. This is an exciting concept that is gaining as rapidly as possible. This is an exciting concept that is gaining is popularly as dumping areas become more and more scarce, but it will be some time before ocean dumping areas reclaimed in the fashion.

Even though we are attempting to utilize our dumping areas for other purposes as rapidly as we can, the concept of zoning still pervades our society. Certain areas are set aside for certain types of use. Society is willing to accept a factory in a particular location where it is hiuden, while at the same time people will not accept that factory as a next door neighbor.

The same concept has been suggested for marine coastal areas. Certain regions would be reserved for industry, others for sewage treatment, while still others would be set aside for recreation. Some of these activities are such that they may be pursued in conjunction with one another, whereas others cannot.

The idea is for every coastal area to try to obtain utilization of marine resources available. In order to accomplish this type of zoning most effectively, a great deal more knowledge, is required than is presently available about the synergistic effects of various types of activities.

One of the major advantage of optimum zoning is that it would allow maximum avoidance of undesirabie synergistic effects. The biggest stumbling block to coastal zoning probably is in overcoming entrenched inteiesss in regions where the optimum activities are different from those presently underway A socio-economic problem rather than a technological one, the Coastal Zone Management Act enco rages coastal states to attempt to take the first step in pl log for optimum utilization of the coastal areas.

After pla are complete, the states area to implement them with the financial assistance of the federal Government. The concept is an interesting one, because it essentially allows some pollution of certain areas while refusing to permit any pollution of other locales. This is somewhat different from the way pollution is presently controlled, where the effort is made to limit pollution in all areas to the minimum possible amount.

With the zoning concept pollution would still be limited as much as possible, but it would be permissible to allow the water quality to degrade below the mean in some places, since in others it would be higher than the mean.

Controlling Pollution by Taxation

One last method of pollution control that has been suggested, although it has met with a great deal of resistance and antagonism, taxation. As is well known, taxation can be utilized to encourage or discourage certain activities in addition to raising funds to support government programs.

Pollution would be allowed, but it would be taxed at a rate proportional to the environmental insult. This general philosophy is followed at the present time by requiring the polluters pay fines roughly proportional to the amount of pollution, and in many cases industries will purposely pollute and pay the fine because it is apparently cheaper than making the necessary plant modification required for cleanup.

Obviously the system is not working if this is the case ; the fines are too low. Some economists suggest that a tax on pollution will simply be a license to pollute and will therefore encourage pollution rather than discourage it. Other economists, however, indicate that pollution taxes should be based on the severity of the pollution with some sort of graudated tax. A little pollution requires a little tax, a large pollution requires a large tax.

An important aspect of this concept is that the words the true and total cost to society of these pollutants. Every time the common resource is utilized by an individual, this society for the right to use this common resource. This payment would be in the form of a tax set at a level determined by how much society economists believe that anyone utilizing the resource should pay for its degraded then the individual degrading should pay for the degradation.

This is an interesting concept that is being discussed in economic circles taxation certainly is a powerful tool for societal behaviour modification that has worked many times. Generally speaking, taxation encourages particular activities and discourages other if the taxation amount is large enough to completely discourage undesirable activities.

From a management point of view there are a number of different directions that can be taken to control pollution, including both technological and socio-economic alternatives. So far society has tenced to aim more in the diection of the technological alternatives, but there appears to be definate swing toward other methods.

In the year to come it appears that more and more emphasis will be placed on these nontechnical methods especially as the tradeoffs, costs, and benefits become better understood.

The Forces Involved in Dispersing Pollutants

Once a pollutant has been introduced into the marine environment, it becomes a matter of no small interest to determine where this pollutant goes and how the concentration of the pollutant varies with time and location.

These process fall into two basic groups : advection and dilution, advection is the mechanism by which the pollutant is carried from one place to another, while dilution results when the pollutant is caused to reduce its concentration by some sort of spreading mechanism. Advection simply has to do with oceanic currents and will include three primary types of current since these are encountered most often in the oceanic regime.

Management of the Marine Environment

Up to this time we have bee primary concerned with the technical aspects of pollution. Space has been devoted to diseussion of what pollution is, where it comes from, and where it goes in the marine environment. Now we must consider how to control this pollution so that the problems resulting from it are kept to a minimum. The previous sentence has a number of words in it which should probably be defined very carefully, but probably never will be.

For example, the statement that pollution should be controlled so that the resulting problems be kept to a minimum may be intrepreted many different ways be many different people since a problem for one group is not necessarily a problem for clear-cut. Just as one man's trash is another man's treasure, one man's pollution is another man's enrichment, and this is not only dichotomy that faces a conscientious marine manager.

There are problems of jurisdiction involved in whether the responsibility for running the operation lies with the federal government, a local government, or some regulatory commission, and always there is the application of priorities supposedly set by society as a whole. The basic tool the manager has to work with is the law, however, and this is where we begin.

The Legal Background

The legal background for pollution control goes back to the Rivers and Harbors Act of 1899, recently interpreted so as to be used as a pollution control measure. The act set up responsibilities for protecting navigational and environmental aspects of all navigable waters to the U.S. Army Corps of Engineers.

The Corps therefore has the responsibility not only of protecting navigable waters so they remain navigable but also protecting them against their loss as a viable vehicle for various other uses that society considers deriable.

An interesting aspect of this law and ope reason it has become such a powerful tool in the hands of the environmental protectors is the redefinition of navigable waters. It turus out that the modern definition includes just about every stream going into a large river in the United States.

Thus the Army Corps of Engineers was given the responsibility for protecting the environment 1899 and to all intents and purposes still retains the primary responsibility as far as brackish and salt water is concerned.

In 1934 the Fish and Wildlife Coordination Act was passed. This has since been amended many times, but essentially it involves the fact the U.S. Fish and Wildlife Service and The National Marine Fishers Services have the responsibility for determining the effect of any proposed project on fish and wildlife resources.

They then report to the Corps of Engineers which makes the final decision as to whether or not the proposed project is allowed. In 1948 the Federal Water Pollution Control Act was passed and this also has been emended periodically. It requires the Corps of Engineers to apply environmental standards and criteria in approving the disposal of spoil in navigable waters.

The most recent amendment of this act was the Federal Pollution Water Control Act of 1972. This was a complete rewrite of the 1948 law and it set out for basic requirements. In the first place, a national goal for the elimination of all pollution from America's waters by 1985 was proclaimed.

This is the famous zero discharge goal Note that this is not a mandate to have zero dischatge by 1985, but as a national goal it is something to aim at and it is to he hoped that we will come as close as possible. Second, the Act of 1972 set as another goal the addition of secondary treatment to all municipal sewage treatment plants by 1977. This goal has not been met but there has been a strong effort in that direction.

Third was the effort to encourage the use of new, more advaneed treatment and disposal methods by 1983 so that some of the problems will be resolved by the use of advanced technology rather than by simply improving yn present methods. And last, the act established an industrial cleanup program with restrictions becoming tighter with the passage of time.

Each year the requirements involving effluent standards will be tightened so that these will be continuing cleanup of the nation's waters. This particular aspect of the bill was backed by stiff penalities including both fines and imprisonments. This 1972 Act is e basic law nnder which most antipollution suits are brought. The Water Pollution Control Act of 1948 encouraged the creation of uniform laws to control pollution, supported research, or firms to cease practices leading to pollution, and established the Federal Water Pollution Control Advisory Board.

This has been succeeded in recent years by the Council Environmental Quality. The year 1965 saw the passage of the Water Quality Act, In this act states were given an opportunity to adopt and enforce federal water quality standards for interstate waters, The Clean Water Restoration Act of 1966 authorized massive federal participation in the construction of sewage treatment plants About 3.5 billion dollars were allocated to be spent between 1967 and 1971.

The National Environmental Policy Act of 1969 provided preparation of environmental impact statements when the U.S. Army Corps of Engineers determined it to be necessary. The Act also declared it to be national policy to encourage the production of enjoyable harmony between man and his environment.

The Water Quality Act of 1970 addressed oil pollution from vessels and offshore facilities, federal permits, sewage pollution from vessels and hazardous substance discharged into the nations's waters.

Also in 1970 recorgnization plan No. 3 was submitted to the Congress by the Chief Executive. This reorganization plan established the Environmental Protection Agency, setting up the organization and transferring responsibility for various activities from other agencies

Unfortunately, the executive order does not address the problems of the objective or the purpose of the Environmental Protection Agency (EPA) and this objective or purpose has been left to the Agency Administrator to interpret for himself. This omission has resulted **in** a general grey area with EPA qeing as strong or as weak as the Administrator and Cengress want it to be.

In 1972, in addition to the amendent to the Water Pollution Control Act, two other act of some import to the protection of the marine environment were passed. The first of these was the ocean Dumping Act of 1972. This act requires the Army Corps of Engineers to apply Environmental Protection Agency standards and criteria in approving ocean disposal of dredge spoil materials where transport passes through United States territorial waters.

This is an interesting law because it controls the material that orginates in United States ports and not that material which is damped in United States water. Consequently no matter where the material is dumped, as long as it orginates in a United States port, the Corps of Engineers has som: control over it.

The other act pass in 1972 having important ramfications to the marine community was the Coastal Zone Management Act. This act gave the individual states primary responsibility for determining land and water uses to more effectively balance environmental protection add economic development objectives in the coastal zone.

It delneated the coastal zone as including all navigable waters and defined them as all waters up to the mean high tide line, including we lands wholly or partially covered at high tide, whether privately or publicly owned. Under this act large amounts of money have been given to participating states, and all coastal and Great Lake states have so far indicated a desire to participate. Fuhds are initially for planning purposes aimed at developing a general zoning plan for the use of the coastal zone Later funds will be available for the implementation of these plans.

In an attempt to get at specific causes of pollution, the Ports and Waterways. Safety Act was passed in 1972 which addressed the problem of pollution from ships.

This act allowed for ship design and construction regulation with respect to possible environmental demage and it applied to all United Stotes flag vessel as well as all foreign vessel entering United States waters carrying either oil or hazardous liquid cargo in bulk.

Vessel esign and operational standards for all these vessel in United Sta es waters were finally promulated on 23 January, 1977. In this the United States government is attempting to prevent as many large catastophies involving vessel carrying oil or harzardous materials as it possibly can by simply not allowing marginal vessel to navigate within the territorial limits of the United States.

Another fact of marine pollution having a source outside the continental area is involved with driliving for oil and gas offshore. The Outer Continental Shel Lands Act gave the responsibility of Granting leases for continental shelf drilling to the Department Interor.

These leases required the leasees to reimburse the federal government for any oil spill cleanup cost and in addition they require the installation of bolwout prevent equipment on the wells themselves.

INTERNATIONAL CONVENANTS

The foregoing is a brief list of the basic framework of federal laws which govern the discharge of waste products into our territorial waters In addition there are a few international convenants under which the United States operates. The Intergovernment Maritime Consultative Organization (IMCO) has sponsored three conventions resulting in a number of regulations.

The first of these was the 1954 Oil pollution convention which was amended in 1962 and 1969. This convention places limitations on the rate of discharge, the oil content, and the distance from land for oily ballast discharge water from ships.

The enforcement of this convention is left up to the flat state, that is the countty where the ship is registered. The second convention was the 1971 amendment to the Oil pollution Convention which branched out in different direction by addressing vessel design. Unfortunately this part of this amendment has not been rectified by any martime nation.

In 1973 a ship pollution convention was sponsored by IMCO, and this convention resulted in a set of regulations establishing standards for ship board sewage treatment equip. ment and als specified the requirement for segregate ballast tanks for new ships of specified sizes. It also specified methods for storing dirty oil ballast residue, oil discharge monitoring systems, and oily water oil extraction systems.

Unfortunately this convention also has not been ratified by any martime nation. Consequently, behaviour on the high seas is left strictly to the conscience of the individual ship's master.

At this time there are also no international agreements on ocean dumping or the environmental effects of seabed resource exploitation. There have been many attempts to draw up an International Law of the Sea agreement but as of this writing none has been successful.

Jurisdictional Responsibility

One of the problems facing a marine resource manager is the problem of jurisdictional dispute. Who does have responsibility '1 In the four general areas the responsibility is somewhat dispersed. The first of these areas is water drainage, capture, or use which is exclusively on a local level. Since this is primarily a problem of private rights which is usually involved with state of private ownership.

This area does sometimes get involved as a pollution

source s: it is of concern to us. The third area of interest includes the bays and coastal waters which are controlled by a combination of state and federal regulations.

Generally speaking, the state has control over resources and the federal government has control over navigational and environmental considerations, but there are many exceptions. Most state laws are not quite as strigent as the federal environmental requirements so that the federal requirements must be obeyed.

However, there are some state laws more strict than the federal and when this is the case, the state laws take precedence. The last areas is the region beyond the coastal region wherein the rasponsibility is somewhat loose. We have seen that there are no viable internetional agreements for the deep ocean, but in many cases the United States has acted unilaterally to protect its continuous waters.

When the passage of the Ports and Waterways Safety Act and the Outer Continental Shelf Lan Act attempts have been made to protect against the effects of oil and hazardous material, corgo ships and offshore oil-wells.

Political Processes

Once the problem of jurisdiction is settled, there then appears the problem of political processes. The political process is such that technical standards will almost always yield to such things as hardship cases, emergency situations, or strong public sentiments.

Consequently, any decision to be made where the environment is involved must have a public input or else the decisi)n will probably not be effective. 'The manager must not only face an often emotional public, but also individual agencies, both state and federal, working at cross purposes.

For example, the Environmental Protection Agency is primarily concerned with water quality, the U.S. Fish and Wildlife Service is primarily concerned with fish habitat improvement, while the Army Corps of Engineers has a primary concern in the areas of navigation improvement.

These concerns must be balanced against common public needs such as high employment, protection of scenic environments, and housing demands, and very often these conflict very strongly so that at least one of the many groups must eventually compromise.

The compromise, in the long run, must be based on some sort of a set of sodietal objectives. Society must somehow or other determine the most important things it wants from its marine resources.

The activities range from aesthetic enjoyment of the area to commercial shipping and waste disposal. Obviously waste disposal is not exactly the type of activity to enhance aesthetic enjoyment, nor is utilizing the coastal zone for military maneuvers encouraging to the maintenance of a reguge or a sanctuary.

Some of the uses of the marine coastal area must be sacrificed in certain cases for other uses which society deems more important. Somehow or other society must set priorities so, that the areas can be managed to obtain the ends that society wants from them. This is a very important point and one that is often completely ignored.

In very few cases has society overtly set down priorities for the use of marine areas, and until priorities are specially stated, marine area managers will be forced into making had decisions. It is to be hoped that some of the planning being done under the Costal Zone Management Act will involve priority setting, especially regarding the use of particular areas.

If this is the only outcome of the coastal zone management act, it will indeed be a valuable contribution to the management of the activities of society in the marine coastal zone.

COSTS, RISKS, AND BENEFITS

In order to set priorities in a rational manner, managers must by one means or another determine benefits, cost, and risks associated with any and all activities. One these three parameters have been quantified, then cost/benefited rations may be determined and priorities may be set on the basis of these ratios.

However, this is much easier said than done. All three of these parameters are usually quite qualitative and very difficult to quantity. Benefits, for example, are often what people perceive them to be and therefore are essentially a political evaluation rather than an exact calculation.

Costs can often be determined accurately in terms of the actual physical nature of the activity. Buildings, machinery, and laboour can be quantified in terms of their cost. However, very often an attempt to control pollution or the effect of pollution on society results in costs that simply cannot easily be quantified.

Risks also are difficult to evaluate. How much is a life worth? Does it make a difference if it is yours or someone else ? Thus we find that we must be satisfied with some approximation to these three parameters. Some methods for obtaining approximations to values for, costs risks and benefits will be discussed below.

The cost versus pollution control curve, is an exponential curve wherein the initial costs of controlling pollution are relatively inexpensive, but the final 5 or 10% of pollution control may cost as much the initial 90 to 95%. The benefit polluation control curve, on e other hand, is parabolic in shape.

Here the benefits in the initial cleanup phase are quite large while the benefits accrued from the final stages are quite small. We might consider then some sort of a cist/ benefit ratio, knowing these two functions.

The optimim cost benefit ration is determined on the basis of the incremental cost per incremental benefit. If example, an additional dollar of environmental control will produce more than a dollar's worth of benefits, then it is more effective to spend additional money on control. If, however, additional control results in fewer dollars worth of benefit, then it appears that the control process has been carried too far.

The optimal level of pollution control is not at point A where the cost of pollution control and benefits are the same, but is at point B, where the rate of increase of cost per unit pollution control is equal to the rate of increase of benefit per unit pollution control. At point B the slopes of the Curves are the same. We may then write

$$\frac{dc}{d(pc)} = \frac{db}{d(pc)}$$

From a cost benefit analysis then it appears that it is not desirable to completely clean up the environment since the benefit accured would not be worth the expenditure required. However, this is not the complete story. It is obvious that in certain cases even a small amount of pollutant is not acceptable because of the danger to health.

Thus the risk too much be considered. Each acitivy and each pollutant has an associated risk, some of these to the environment and some to man. It is difficult to quantify these but one method commonly used is to deter mine the number of fatalities per million hours of exposure for each

Table 8.1: Risk Table.

Risk	*Death rate per millionhours of exposure*
Voluntary	40.0
Rock climbing	6.6
Motorcycle	2.4
Scheduled airline	2.4
Smoking cigarettes	1.2
Death by diseaseIncluding old age	1.0
Private auto	0.95
Sking	0.71
Railroad and bus	0.08
InvoluntaryElectric power	0.002

person involved in the activity considered. This is expressed as a statistical probability. Table elsewhere in this chapter is a listing of risks expressed in this manner for various common activities.

Notice that the activities are divided into two groups, voluntary and involuntary activities. The voluntary activities are those which the individual can decide for himself whether or not he wishes to do. Notice that many of these activities have a probability very close to that of death by disease, including old age.

Others, however, such as rock climbing, have an extremely high risk associated with the them and therefore are only for the adventuresome.

Interestingly enough, the risk associated with the one involuntary activity listed, electric power, is less by about three orders of magnitude than most of the risks in the voluntary portion of the table.

Apparently then we are loath to let others do in to us what we happily do to ourselves. Pollution of course, falls within this ar a of involuntary activities since any pollutant discharged to the arine environment is not directly controllable by any individual sequently we would expect that the risks associated with pollutants would be down in the area of that of electric power rather than up with that of railroad and bus travel.

In order to quantify priorities, it if necessry then to develop some sort of risk benefit ratio, which in trun, requires quantification of benefits.

One possibiliy of quantifying benefits is to simply assume that in the case of voluntary activities the amount of money spent on the activity by the average involved individual was proportional to its benefit, while in the case of involuntary activities, the contribution of the activity to the individual's annual income as porportional to its benefit.

Here risk is plotted relative to the benefit for various kinds of voluntary and involuntry exposure. From these graphs and some other works that has been done, a number of conclusions can be di awn. First, the public is apparently willing to accept voluntary risks roughtly a thousand times greater thnn those involved with involuntary exposure.

Second, the statistical risk of death from disease and old age appears to be a psychological yardstick for establishing the level of acceptability of other risks. Third, the acceptability of risks appear to be crudly proportional to the third power of the benefits, whether they be real or imagined.

Fourth, the social acceptance of risks appears to be directly influenced by public awareness of the benefits of an activity as determined perhaps by advertising or the number of enhanced by increased public awareness. A determination of the level of optimum pollution control and general management schemes involved with addressing the total pollution problem is not as easy one.

We find that everything costs something. There is no such thing as a free lunch For every cleanup process involved there is a cost. Sometimes this cost is minimal but other times the cost is extremely high. For example, it has been estimated that in order to meet the 1983 goals of the Federal Water Pollution Act or 1972 it would cost 350 billion dollars just to treast municipal sewage in an adequate manner.

This includes the cost of controlling nonpoint sources along with sewage treatment plant updating. Whether or not the America public is willing to spend this much on this sort of program is at this time unclear, but the American public must be madea ware of the necessity of costs and trades of any cleanup program.

If we are going to spend this amount of money for pollution cleanup then there will not be as much money available for other activities, and once again we must decide which to these activities is more important to us.

But the decision can only be made if we have as much data as possible in hand to allow as accurate as possible a determination of benefits, costs, and risk involved with each process.

THE FUTURE OF THE MARINE ENVIRONMENT

In the previous pages we have seen that the oceanic volume is extremely large, while at the same time, that portion of the oceanic volume available for pollutant discharge is relatively small. Consequently, we hope to have demonstrated a need for some kind of marine pollution management since there is a very real problem associated with dumping wastes in the ocean.

It has also been shown that an ever increasing population, distributed primarly around coastal areas, will inevitably incrcase the total amount of waste discharged to the oceans. This is coupled with the fact that technology is continually growing and its growth will be associated with an increasing amount of pollution per capita with the passage of time.

It all boils down to the realization that the total volume of water available for waste dislution will remain relatively constant while the total amount of waste will probably increase as far into the future as can be foreseen at this time.

Management Requirements

Fortunately, however, in those nations which have reached a relatively high level of technology, there is an increasing awareness of the problem and a public requirement that planing and management proceed in order to utilize the marine resource in the best man er possible. In order to do this a number of requirements must be met.

In the first place the effects and pathways of all pollutants dumped into the ocean must be known. This is an extremely difficult task. Many of the patways are known but the effects are not. For other pollutants the effects are well known but the pathways back to man are still mysterious.

This obviously will require a great deal of research, and even then there is no guarantee that all effects and pathways for all pollutant materials will ever be known. Second, the average residence time for all materials discharged into the marine environment should be known.

Here again we are faced with an almost impossible problem due to the tremendous amounts and number of different materials discharged into the oceanic environment.

There should also be a continuing monitoring of the environment if only to determine gross changes so some action may be taken before the situation gets out of hand.

Lastly, we must continually determine and redetermine what people consider to be important. The more and folk ways of society constantly change, altering values and priorities.

Setting Priorities

At this time managers have no clear understanding as to what people's priorities are, nor is there any real indication that the people themselves have any great insight into their needs and desires. We do not know, for example, whether the major portion of the populace considers sport fishing or ocean dumping to be higher in priority with respect to usage of the marine environment.

We do know what the small special interest groups have to say about this situation, but in terms of jobs lost or consumer goods price increases which would result from the discontinuance of ocean dumping, there is very little information. In order to determine society's needs and desires the people must first be educated as to the costs, benefits, and risks associated with any environmental protection solution being considered.

Obviously a much better choice of alternatives can be made if the advantages and disadvantages may very well be a utopian dream, but it certainly should be the goal of each manager to come as close to this as possible.

At the present time choices are made on the basis of limited kuowledge since not all of the costs, risks, and benefits are availabe to the manager making a decision, every those that are available to the manager are usually not available to the pulbi at large so that the manager's decision can be evaluated. This situation must change if utter choice is to be avoided.

Once our cards are on the table where they may be seen by all interested parties, we may proceed by setting goals that are both popular and attainable. A goal such as "zero discharge by 1985" is not an attainable one and when such a goal is set and not met, very often the result is a certain cynicism regarding all future goals.

Goals, Guidelines, and Standards

At this point it may be desirable to define some words commonly used in the management of the marine environment. In particular there are four words needing clarification: goal, guideline, criterion, and standard. Although the definitions which follow may not necessarily agree with those in a standard dictionary, they follow the usage of most people involved in marine are management, standard: A plan established by governmental authority as a program for water pollution prevention and abatement.

Criterion: A scientific reuirement on which a cecision or judgement may be based concerning the suitability of water quality to support a designated use.

Guideline: An acceptable methodolgy for achieving any given standard.

Goal: The ultimate standard which all current standards may oventually approach.

Involving the Public

Thus goals must be set with two criteria in it ind. One is the scientific requirement defined above, but there is also a personal requirement on the part of the public based on its sense of priority. The public will not support a particular criterion, even with substantiating scientific data, if it interferes markedly with the present lifestyle.

Until the citizenry can be educated to the risks involved in not accepting scientifically based criteria, looser criteria must surface. But it is the duty of the manager not only to choose the acceptable criteria but also to attempt to educate so that more logical criteria may receive public support.

Along with the choice of particular goals, standards, criteria, and guidelines for a particular time, it must be assumed that these are not permanent but may very well change with the passage of time. Both the environment and public attitudes must be constantly monitored to effectively meet the needs of a changing society.

With a proper monitoring system, not only may the standards, criteria, and guidelines be change but also the goals may update. In this manner the public is made a working part of the management process adding support to the process and consequently making it more effective.

The coastal Zone Management Act was designed along these line and it has been in operation long enough for a major portion of coastal states to have prepared detailed plans on the optimum use of coastal areas.

These plans are undergoing extensive public hearings and it is to be hoped that when they are finally adopted the public will actively support them. If thib is indeed the case, it can be hoped that the effect will, at the very least, be to stop the increase in marine pollution and may be even reverse the trend by decreasing pollution.

In any event, however, regulations set down to protect the environmental quality of the marine environment will be enforced with popular support.

IMPACTS ASSOCIATED WITH CLEANUP TECHNIQUES

Cleanup Effects on Marshes

The equipment used for several cleanup techniques considered here does not impact on the marsh itself. For example, mechanical and weir skimmer are deployed in tidal channels, and pools, not in the marsh proper. Vacuum skimming in itself produce no impact on the marsh, but it is frequently accompanied by substantial foot traffic.

Fool and equipment traffic and considered secondary

impacts and may, in themselves, produce substantial adverse effects. The following sections are modified from Maiero et al. (1978).

Effects of Traffic

The most obvious type of disturbance caused by traffic in a marsh is physical breakage of plants. In both grass dominated and succulent-dominated marshed, physical damage effectively reduces the amount of photosynthetic tissue and may expose the interior of the plants to toxic franctions of the oil.

Plants are likely to recover from one such event, but recurrent trampling may clear a parth that will persist for years. If the soil is soft, as it is in many marshes, roots and rhizomes maybe broken and thereby accelerate erosive processes.

Foot traffic potentially accelerates erosion even where the substratum is firm and the rhizome may remains essentially intact. Shorelines that consist of steep escapements are particularly vuluerable. Cleanup activities that entail foot traffic should be used in such regions only after other methodologies have been considered and rejected.

Sometimes it is necessary to transport heavy equipment to remote portions of marsh. The impacts of foot traffic described above apply even more strongly to vehicular traffic. Traffic on soft marsh soil may bury plant stems and leaves and reduce their productivity until they resprout or grow back above the soil surface.

Further, traffic under these conditions is likely to bury oil. In the anaerobicsoils that characterize mangroves and marshes, residual oil may persit for years. If the buried oil is toxic, it may inhibit the growth of anything in the contaminated zone until it eventually dissipates.

Flushing

Low pressure flushing with sea water (from a nearby source and thus likely to be of a salinity to which the marsh is accustomed) may be beneficial if used with caution.

Where vegetation cover is continuous and sediments relatively stable, low pressure flushing may be effective in removing substantial amounts of oil from vegetation.

Drawbacks of the method are

— All oil is not removed from leaves and a sufficient amount may adhere to the waxy plant culticles to cause damage. This is especially true when the oil is fresh in which case it adsorbs onto leaves very strongly an penetrates them.

— Foot traffic required to deploy the flushing equipment may cause damage.

— Oil flushed to creeks, etc., will be available to re-oil the marsh or other areas unless properly collected or otherwise removed.

Where vegetation cover is incomplete where much bare mud is in evident, or where sediments are sort and muddy, sediment disturbance or erosion may produce additional damage to that of the oil. Wherever possible, sediment disturbance should be avoided.

High pressure flushing may cause some erosion, local rearrangement of the substrate, and physical damage to the plants. These forms of damage may be less severe than similar impacts caused by foot traffic, but this depends upon mode of deployment.

High pressure flushing may cause oil to be driven into the substrate. If steam or heated water is used, marsh animals and plants in the spray pattern may be killed or stressed by thermal stock.

Sorbents : Sorbent pads, oil snares, and similar cleanup aids have two major drawbacks

— They are usually used by a large group of personnel who heavily traffic the marsh and

— They must be recovered. Additionally, if cleanup teams are not trained in the use of these material, there is chance that oil may be mised with a shallow layer of the substrate in the cauree of recovery.

Because complete recovery of sorbent materials is seldom possible in a field cleanup exercise, remanants of sorbents may persist. Most remanants are merely unsightly, although some may ensure birds and animals. Large concentrations of undergraded materials could block the ligh from an appreciable portion of the marsh surface and consequently reduce marsh productivity in that region. Biodegradable sorbents avoid many of these problems.

Oil sorbent materials (synthetic and biodegradable) should always be recovered from the marsh, those that are not collected are a potential source of recontamination and a hazard to marsh animals.

Cutting

Cutting of oiled plants is a cleanup technique that entails direct physical destruction of plant tissues. As such, it severely reduces the amount of photosynthetic tissue and may expose the interior of the plant to toxic substances in the oil. Moreover, cutting operations are likely to require a great deal of foot traffic and the attendant adverse effects.

Cutting is probably most beneficial with certain species of grasses and rushes that have been heavily contaminated with viscous oil and are not subject to natural cleaning. If the entire aerial portion of the plant is coated, the roots are likely to suffocate unless some passage way is opened to the interior of the plant.

Cutting accomplishes this, provided that free oil that might replug the cut stem has been removed from the surrounding marsh. However, if oiled marsh plants are not throughly coated with oil air can still diffuse down the stem to the foots, so that cutting is unnecessary unless other threats are present.

Spartina marshes are very tolerant of occasional cutting, especially late in the growing season. Saltwort marshes are less tolerent. Burning is sometimes an effective method of removing oil and contaminated vegetation from a marsh without encouraging injurious foot traffic.

Spartina marshes can withstand occasional burning. In

fall and winter, they die back to a state of dormancy. During this period, the plants are dry and may support burning. In fact, fall burning of marshes used for agriculture is a commonly applied management tool.

During the dieback period, burning can be achieved in spartina marshed without damaging the buried portions of the plants and can stimulate their regrowth. In all other seasons, however, not only in Spartina difficult to burn, but the growning portions, shoots, and buds are injured by burning.

However, this damages does not necessarily permanently harm the marsh since the underground portions of the buried plants are likely to sprout and replace the destroyed portions soon after the burning. Saltwort marshed do not die back seasonally and the plants do not have large, proteted underground systems ; thus, burning is injurious at any time.

Soil Removal

Removal of soil entails elimination of marsh plant habitat and should be avoided. Nonetheless, if the substrate is heavily saturated with toxic oil and no predicted to recover naturally, this may be the only available cleanup technique. On soft areas of find mud, it may not be feasible at all.

Dispresent Chemicals

Dispresent chemicals have been applied to oil in salt marshes following spills and in a number of experimental situations. In many cases, these have caused additional damage to the oil alone, and in some cases can be demonstrated to enhance the penetration of oil into intertidal sediments.

Recent evidence suggests that new generations of dispersant may cause little additional teological effects to oil along when used at appropriate concentrations. However, the value of using them in most cases if unclear particularly on extensive marshes where, following application, dispersed oil may enter creeks to affect other part of the marsh or intertidal systems

The may be of value in cleaning small and stable fringing marshed in combination with low pressure flushing, but, in general, dispersant use is not recommended for salt marshes. It should be borne in mind that spraying of near dispersants onto vegetation products damage and correct dilution to manufacturer recommendations is vital.

Marsh cleaning methods do not appear to decrease damage done by oil and often increas in. The need for cleanup should be carefully reviewed before any action is taken.

Cleanup Effects on Mangroves

Few publications are available that deal directly with the effects of oil spit cleanup on mangroves. The several observations of the effects cleanup on mangroves at the Florida Keys oil spill of 1975. So many authors summarize effects of cleanup of marine wetlands and include numerious general statements concerning mangroves.

The following summary is based on the publications and their personal observations several scientists made during cleanup operations within oiled mangroves.

Recovery of Marshes from Oil Damage

Although a definition of recovery has been attempted, mainly to distinguish the process from that of restoration, the definitive includes the terms structure and function. The majority of scientific literature on the effects of oil deal with changes in the structure of marshes relatively few deal with the function of marshes let alone the part oil damage may play.

Hence, we may be able to determine the time at which a marsh returns to something approaching its original structure, but not necessary. Alternatively, a similar marsh function could be achieved with a different structure. This is a line of investigation that desrves further attention, but at present we can do little but recommend that marsh structure be used in the assessment of recovery.

In an earlier section, the rate of recovery of salt marshes has been shown to be considerably influenced by a large number of factors Their precise action and interactions are often poorly known, which means that in practice each spill must be treated as unique and assessment of damage must be dealt with in each case on its own merits little more than generalization is possible from the scientific literature.

Recovery processes begin when oil toxicity is reduced or removed. In most cases, recovery starts irrespective of the activities of man, although cleanup or other action may influcence the rate.

Where oil damage is relatively light and little of the toxic material remains in the system recovery can be rapid either by regrowth from rootstock or from seeds.

Baker (1971 a and personal, communication) reports that single oiling in field experiments, although toxic to mary salt marsh species, may produce effects detectable only for one to three growing seasons.

Where oiling is repeated or where damage is servere and toxic materials are retained in sediments, recovery may take much longer, and the effects of the spill may be detectable over decades. Where erosion of sediments taks place, recovery may not occur.

OIL SPILL PREVENTAION AND CLEANUP IN THE VICNITY OF CORAL REEFS

Prespill Mapping

Prespill contingency planning is now an accepted part of prepartion for, and response to, oil spills. Vulnerability mapping has been used by several authors to identify those area of shoreline that may be particularly sensstive to oil spill damage and to help ensure that damage due to the use of inappropriate cleanup procedures is minimized.

One frequently used vulnerability index was devised by using the assumption that coastal geomorphology frequently determines the types of econological communities found there.

Information about points of access to a spill site is

essential in the event that men and equipment must be deployed at shore notice.

Therefore, accessibility mapping is an important adjoint to vulnerability mapping. Coral reef vulnerability to oil spill damage will vary with reef type, zonation patterns, and tidal actions For expmple, in a fringing reef, the seward facing reef creast is generally exposed to high wave stress ; here oil will probably have a short residence time before being dispersed naturally.

In contrast, wave action is a lot less severe in the shallow reef flat that frequently supports seagress beds. Here, a long oil residence time with a correspondingly high risk of oil damage is likely. Similarly, oil that strands on reefs with a limited tidal range will likely tend to persist for longer periods than that in areas with large tidal ranges.

Although vulnerability and accessibility mapping techniques are now well established, little evidence suggests that such techniques are being applied to reef ecosystems. Some reef areas are well mapped and information is readily available on the location of refineries, tanker terminals, oil platforms, and major tanker routes.

This information can be used to produce maps that help locate areas of high potential risk. Newly developed techniques for mapping large areas of coastline, such as low altitude video overflights seem particularly suitable for use in reef areas but have yet to be applied.

Cleanup and Treatment of Spilled Oil

The sensitivity of reef ecosystems and the likelihood of their being impacted by oil will depend upon a variety of factors, including the quantity and type of oil and the degree of weathering. The likelihood of cconological impact will also very according to whether the feef is an emergent, shallow submergent, or a deep water type The feed and laboratory studies reviewes indicate that the likelihood of damage is increased if the oil is is incorporated into the water column of if it comes into direct contact with the coral surface.

This, therefore, places the margent or shallow submergent section of a reef as areas of high potential sensitivity. It also indicates that any cleanup of treatment attempt should endeavour not to enhance the transport of oil into the water column or into the permanently subemerged sections of a reef.

Careful consideration should be given to the treatment of oil in the vicinity of a reef as situation may develop where the cleanup or recovery attempt may be more damaging then the oil alone. Baker (1970) has pointed out that merely the shipping activity associated with the cleanup attempt may be damaging due to sediment resuspension by propellers and anchor drag breaking up corals.

Although practical experiences of oil cleanup in the vicinity of coral reefs are lacking, mechanized techniques for containment and recovery of oil using booms and skimers are acceptable as they do not cause oil to sink or become incorporated into the water column.

Any restriction on use of these techniques in coral areas is likely to be brough about not by the risk of ecological damage but other environmental considerations such as high current speed, servere wave action, or shallow water depth that make their deployment impossible.

At present, only few data are available on the impact of disp rsants and dispersed oil on corals. Most information that is available has been derived from laboratory studies.

Laboratory studies have also been carried out on biota associated with reefs. Both categories of study noted damage by dispersed oil, but the applicability of these in vitro data to a field situation is questionable. Similarly, these studies used early generation dispersants which of themselves, where toxic to the carats.

Using dispersants currently, available, indicated that concentrations of 20 to 50 ppm dispered oil in a 24-hours exposure produced temporary stress reactions in corals. Few deleterious effects from 24-hour exposures at 1 to 5 ppm wgrq observed.

An oil removal technique that cannot be recommended for use in coral arsas is the application of particulate sinking agents. These materials absorb oil and because they have a density higher than that of sea water, sink with the oil attached. Deleterious effects could be expected from the sunken oil coming into direct contact with the corals.

Several methods currently in the developmental stage appear to b! potentially suitable for use in coral areas and other marine environments likely to be sensitive to oil. Among these are agents that have been developed that, when added to oil, gel in a semisolid from that can then be recovered.

Research is also being carried out to develop nutrients that accelerate the bacterial biodegradation of oil. Several studies have shown that corals are seemingly uneffected by a layer of oil floating on the water surface.

Therefore, where unsuitable hydrographic conditions or problems of access preclude the use of mechnical containment and recovery techniques, it is probably best, given our present knowledge, to leave the oil alone to weather naturally.

Measures of Damage at the Species Level

Structural changes occur on a reef when a species or combination of species that are dominant components of a system are replaced by other species are eliminated from the system. If the replacement of elimination of a species is significant, then community, or functional level changes can become evident.

Changes in species richness, coral cover, and species diversity are potential criteria for the assessment of catastrophic structural demage to corals and any subsequent recover. Loya (1972) has modified the Shannon-Weaver diversity measure for use with corals as well as any encrusting assemblage.

Grigg and Maragos (1974) used these parameters to describe recolonization by hermatypes on lava flows, as did Sheppard (1980) when assessing the impact of harbor construction on corals. *Porter et al.* (1981) used species richness, coral cover, and relative abundance to asses structural recovery of hermatypic corals following hurricane damage.

In a 'recent paper, pearson (1981) argued that measures of diversity, percentage cover, and similarity taken individually do not provide useful measures of either damage as recovery.

He suggested that measures of damage and recovery should totally incoroporate measures of percentage cover, mean colony-height, surface-index, species diversity, and simlarity indices. In addition, because reefs are dynamic systems that can undergo change due to numerious environmental factors, any observed structural changes should be evaluated in terms of the structural charactcristics of neighbouring reef systems.

Measures of Damage at the Community-Level

The structural diversity of reef ecosystems is paralleled by their function a complexity. When attempting to assess damage and recovery of coral reefs in functional terms, to useful line of investigation may be development of indices that provide a measure of total reef functions or malfunction.

One major functional characteristic of reef ecosystems is their high rate of primary production that may be 10 to 100 times greater than that of the phytoplankton in surrounding waters (Lewis 1981). Much of this reef primary production is attributable to benthic alage.

A principal determinant of this high rate of productivity in nutrient poor areas is the retention and recycling of nutrients within the reef. This feature may be a useful means of determining the status of total reef function.

For exampie, Pilson and Betzer (1973) found that, although concentrations of phosphates in water passing over a reef were low, there was no difference in the upstream and downstrem levels, thereby indicating that a reef recycling mechanism was present. Evidence of phosphate recycling has been found by Pomeroy et al. (1974).

There is also evidence of ammonium and nitrate recycling mechanisms. An increase in nutrient loss suggestive of a breakdown in nutrient recycling mechanisms may be a potential indicator of a total reef malfunction. Further evidence for the steady state nature of a reef ecosystem is provided by the rati) of gross production to community respiration (P/R) that frequently approaches unity (Lewis 1981). Again, any large be change in P/R ration could be indicative of reef malfunction.

Coral Reef Recovery

Our ability to measure recovery in a coral reef system hinges on the success of a program to define or measure than is damage. Endean (1977) said that complete recovery "involves the return of those species typically associated with corals and forming part of the coral reef community in the particular geographical region concerned and the reestablishment of the complex relationships normally existing among the species,"

While this definition is theoretically correct, from a practical sense in terms of measuring recovery, it is probably too broad. Coral reefs are among the most complex marine ecosystems in terms of communities of structure and ecological function. Monitoring all aspects of reef biology in an attempt to confirm a recovery or to define a stage of recovery is not feasible.

There are, however, aspects of reef structure and function that could logically be used as indicators of a level of recovery. Specifically, a coral reef has patterns of species composition, abundadce, dominance, reproduction, recruitment, growth, and mortality among the hormatypic organisms.

A gauge of recovery may be generated by comparing qualitative observations and quantitative measurments of these patterns on the recovering reef with those presumed to be characteristic of the chosen model of a fully recovered reef at the site in question. This methodology, however, requires the assumption that recovery of the coral organisms will eventually result in the recovery of the entire reef community.

Few reefs in the world have been subjected to quantitative ecological assessments. The means that it is unlikely that any site-specific information will be available for comprative purposes.

However, reef morphology and biotic community structure are typically similar over a definable range of variation within the area or region. Thereloi e, there is likely to be a basis upon which to roughly estimate the former structure and condition of nearby unimpaeted reefs existing under similar conditions of natural environment, along with the remains of the impacted reef.

If it is possible to infer preimpacted structure from regional patterns, there is at least a structural and functional model for a recovered reef at the locality in question.

The measurement of recovery is complicated by the fact that recovery on coral reefs tends to be a long-term process Endean (1977) found that the recovery period is correlated with the extent of the damages inflicted on the hard coral cover.

Small scale localized destruction from natural events generally requires less than 10 years for recovery provided that major sectors of the community remain virtually intact and the area is conductive to coral growth. Heavy destruction requires 10 to 20 years for full recovery, while especially severe impacts may require several decades for complete recovery.

If a chronic source of pollution is present in the area of the damaged reef, recovery may be further prolonged or may not occur at all (e.g., see Rinkevich and Loya (1977).

Cleanup

What cleanup measures are appropriate and what are their relative effectiveness are critical questions regarding oil spills in tundra and taiga.

Whenever possible, hydrocarbon spills should be avoided or their effects mitigated. Several measures can be employed to help accomplish this.

All above ground fuel storage containers should be contained within a pit or depression lined with a hydrocarbons in previocs layer. Oil pipelines should be designed to withstand weathering, corrosion, fnd limited displacements such as those due to earthquakes.

Finally, whenever possible, oil pipelines should be routed to F (void sensitive ecological stress such as populations of endana tiered species. Unfortunately, containment or evoidance is not always possible, so cleanup measures for the various soil and a vegetation types must be considered.

Once a spill has occurred, choice of cleanup strategies ill depend upon the type of oil spilled, the nature of the pill the amount of oil, the season of the spill, and the location of the spill. Site aspect, vegetation, and soil types are critically mportant factors.

In addition, the presence of permafrost in ost areas of tundra and much of the taiga will preclude the se of many cleanup methods routinely used in warmer limates or nonpermafrost areas. For example, in permafrost areas with a high ice content, any surface disturbance, such as removal of the oil-soaked prganic soil surface horizon, tillage, or vehicle movement, will nduce thermal degradation of the permafrost and subsidence thermokarst), which will increase rather than decrease total Environmental damage of the site.

Under such ponditions, passive (or no active) cleanup should be considered. Passive measures may also be the pest means of facilitating natural recovery in cases of small pills of crude oil or contained or limited spills of industrial Ails with high proportions of light fractions, much of which will be readily lost by volatilization.

However, recent work by Linkins and Feteher has shown that residual oil in the Oe-01 horizon of tussock tundra soils may alter reproductive and biomass allocation patterns of Betula nana and Eriphorium vaginatum such that their successful reestablishment in oil-contaminated soils is questionable.

When active cleanup measures are necessary, there is heavy reliance on site access for transporting equipment and manpower to the spill. In remove locations, this can involve considerable cost as well as be environmentally damaging unless it can be accomplished with the appropriate equipment, especially during the persnow thawed soil times of the year or snow preseason.

Generally, the choice of which active cleanup techniques are used depends upon the time of year when cleanup must occur. In winter periods when the ground is frozen or snow covered both access and cleanup are facilitated.

The dormant stage of the plants and low winter temperatures that aid in cooling and congealing crude oil so that its spread is limited both contribute to minimizing damage. Snow also acts as an absorbent and cushion protecting surrounding vegetation from oil exposure as well as vehicle damage.

Generally, the oil-contaminated snow can be mechanically scraped off the site, removed, and placed in properly contained areas such as was done at the February 1978 Steele Creek spill along TAPS. Burning the spilled hydrocarbon may also be used as a cleanup measure either before or after mechanical removal.

Combustion may be difficult to initiat and generally will be incomplete. However, burning can help to remove the more toxic light fractions of the oil if done soon after the spill occurs. Smill spills may be difficult to ignite if not immediately burned, but large spills seem to burn readily even several months later.

This may be due to a large proportion of the light fractions being rapped withing the soil in large spills as surface weathering produces as extensive asphalitic cover. Burning is relatively inexpensive, may leave a tarry residue on the surface that can inhibit microbial decomposition of the oil and vegetative recovery. Finally, following any winter cleanup procedure, appropriate measures should be taken to ensure against, remobiliza tion of the oil during snow melt.

Such measures as absorbent booms, straw, or in nonpermafrost and thaw stable permafrost areas, containment ditches or dikes may be used.

The hydrocarbon will then be intercepted or contained to facilitate remov al. The choice of active cleanup measures during the summer or snowfree period will depend upon the phenological stage of the plant, the soil moisture content, depth of the soil water table, and ice content of the permafrost as previously discussed.

Mechanical removal of oil contaminated soil should be considered only at times when potential damage to plants is great from prolonged extensive contact with the hydrocarbon, the ice content of the permafrost if low, and there is relatively easy accesss to the spill site.

Also, the volume of the spill should be such that natural recovery is unlikely and there is high potential for extensive contamination of adjacent areas through contained movement or later remobilization.

If removal of the contaminated soil is not possible, burial can be considered. However, burial should be done only in thermally stable, frozen material since subterranean remobilization can occur and cause contamination of adjacent areas.

Oil may also be removed by floating the oil and vacuming it off the surface of the water. This was used successfully at the TAPS value 7 spill in July 1977, ready access to the site as well as availability of a large volume of water, a wet site, and feasibility of containment.

However, in some cases, oil may contin·ie to appear on the surface of the water for a year of more after the spill ai it is displaced from the surface organic horizon.

Finally, manual cleanup with minimal soil disturbance should be considered in permafrost areas. Absorbents such as straw or commercial materials can remove much of the oil present on the ground surface. Although costly and very time-consuming, thls method will minimize any physical disturbance and decrease the chance of thernai degradation of the spill and adjacent sites.

The extensive reliance on reseeding with non-native species and long-term fertilization, whereas a relatively easy but expensive means of revegetation, does not guarantee optimal long-term revegetation or restoration.

Observations of natural restoration on a 1949 disturbance and on a 1970 disturbance at Barrow, Alaska suggest that it way be preferable to let some hydrocarbon disturbances recover naturally through reinvasion by native plants in lies of reseeding with exotic plants.

Recent work discussed by Shaver et al. (1983) that shows that the shallow organic soil horizons can serve as a major seed bank for revegetation suggests that preservation of the upper organic soil horizons may also be preferable if possible to maximize potential for native plant reestablishment.

Their discussion also points out the possible deterrents of fartilization on delaying restoration of the natural vesetatiort. Penetration of the hydrocarbons into soil plant rooting zone as previously discussed will influence the choice of revegetation/restoration techniques.

On wet sites where there is limited penetration of the hydrocarbon into the soil rooting zone, natural recovery by the native vegetation may be sufficient if cleanup activities adequately remove the surface hydrocarbon contamination.

On drier sites, however where the hydrocarbon will rapidly penetrate the soil rooting zones, killing the majority of the vegetation, it flay be necessary to institute extensive soil cleanup, rapid revegetation efforts.

This will be especially true if site stability (othermokarst, slope erosion, etc.) is a problem. Regardless of the cleanup, revegetation/restoration activities it should be reiterated that care should be taken to preserve of site integrity. Slower spill site revegetation/restoration should always be preferable to risking both on and off site degradation.

Hydrocarbons spills in the Arctic and subarctic cause both short-and long-term effects on the soils microbial populations, vegetation, and wildlife. Although a number of descriptive studies on the effcets of hydrocarbon spills have been conducted, these have generally focused on short-term effects.

Furthermore, thcre has been very limited development in our understanding of the important mechanisms of the direct toxic effeets of oil and the effects of altered soil characteristics on organisms.

When our ignorance in these areas is compounded by our limited knowledge of the in site rates and nature of hydrocarbon degradation in cold soils, it makes it very difficult to estimate effectively the duration of the influence of hydrocarbon in the soil.

Likewise, it is very difficult to predict which cleanup, revegetation techniques will be most successful in alleviating hydrocarbons related deterrents to microbial population and vegetation growth.

Development of better methods for restoring Arctic and subarctic hydrocarbon spills requires a commitment to longterm research efforts focusing on integrated cleanup, revegetation/restoration research.

Research should focus on the development of integrated postspill functionally based ecosystem efforts to determine and evaluate the relative importance of factors limiting site restoration.

These studies should be connected on new sites that have been throughly described and on existing spills sites where adequate documentation exists as to the specific cleanup and revegetation activities that were employed.

Efforts should also focus on natural hydrocarbon seep areas and long-term unaided spill sites where native vegetation exists. Integrated research efforts directed toward these goals as well as toward associated off site cleanup and revegetation disturbances should then begin to provide the information necessary to formulate effective revegetation/ restoration programs in Arctic and subarctic areas.

Unfortunately, until research is initiated, there will be limited advancement in more effective, less costly revegetation/restoration practices.

Chapter 9

PELAGIC FOOD WEB

Deep-sea animals are most diverse under the surface waters of the subtropical and tropical belts. Here are clear blue waters, where photosynthesis by the microscopic algae of the plankton is possible down to a depth of 100 metres or more.

Apart from the divergent regions of the equatorial current systems, where upwelled water brings nutrient salts to the surface and leads to relatively high levels of productivity, these warm central oceanic waters are low in nutrients and standing crops of plankton.

Blue-green algae thrive in such surroundings and a dominant form, *Trichodesmium,* whose rust-coloured blooms may at times cover wide stretches of the warm ocean, is reported to use fixed molecular nitrogen for its growth.

In other respects the composition of the phytoplankton of the tropical and subtropical belts is much like that of other oceanic regions. But the size range of the algae extends below that of floras in nutrient-rich waters, such as are found in the antarctic and in upwelling waters off Peru and elsewhere. For comparable forms the smaller the cell the less are its nutrient requirements, but the more its surface for nutrient absorption compared to its volume.

Thus, it somehow seems right that much of the flora of nutrient-poor tropical waters should consist of mobile I&-

flagellates, such as cryptomonads and chrysophytes. To be but a few microns in size confers another advantage. In subtropical and tropical regions in particular, it seems that the multiplication of the phytoplankton barely keeps pace with its consumption by all manner of herbivores in the zooplankton.

One might then expect selection pressure to be high for, means of countering the grazing inroads of zooplankters. To be small enough to pass through the meshes of a filtering system, particularly of copepods, is advantageous, though there is less escape from the mucus nets of the pelagic tunicates.

But in overall terms of nutrition W-flagellates may form no more than a small part of the grazing of herbivores. For instance, during an expedition to the eastern tropical Pacific, counts of phytoplankton constituents were made (under an inverted microscope) from 200 samples covering 48 stations.

When eaeh count was expressed in terms of its carbon content, the main conclusion was that though flagellates (2-5μ in diameter) formed about 80 per cent of the catches in numbers of individuals, they made only a minor contribution to the total content of phytoplankton carbon.

In equatorial waters most of this carbon was contained in the diatoms, whereas dinoflagellates weze the main carbon contributors north and south of the equatorial region. No doubt the relative amounts of the major groups of algae change from time to time, and we must always remember that the fragile and elusive it-flagellates are likely to be underestimated.

There is no off season in the production of phytoplankton in warm oceanic waters: throughout the year diverse zooplankters live on thinly dispersed craps particularly in the subtropical regions bounded by the gyres. In temperate and cold regions there is rich grazing in the favourable season, during which animals form stores of fats and oils to help tide them over the winter months.

Whatever the place, there are times when the phytoplanankton crops are so meagre and thinly spread that one may wonder how the herbivores manage to exist. Indeed, such thoughts, stimulated by the quantitative studies of Hensen, Lohmann and Brandt towards measuring the productivity of the seas, led Putter to investigate the nutritional requirements of marine animals (as measured by their rates of earbon dioxide production).

After comparing the metabolic needs of diverse species with ambient amounts of algal food, Putter concluded that marine animals must depend largely on dissolved organic matter, which they absorb through their integument.

Krogh, who was drawn to this problem through his experiments to determine whether aquatic animals are able to absorb organic substances, first concentrated on an accurate me od measuring dissolved organic matter (DOM) in sea water.

Though he found that there is much more DOM than particulate forms of organic matter in the ocean, he also observed that DOM is very largely resistant to bacterial decomposition. Krogh concluded that the food of aquatic animals is generally in the form of organisms and organic detritus.

Since the 1950s numerous experiments on diverse aquatic inverte-brates have shown that most marine, but not freshwater, forms can take up glucose and amino-acids through their integument.

But Korgh's conclusion still stands. Even the Pogonophora, which are confined to sediments rich in organic matter, have a restricted rate of absorption when in their tubes.

Members of the micro-zooplankton, certain copepods and euphausids, shelled pteropods and pelagic tunicates form the main groups of oceanic herbivores. More than we realize, the lives of these diverse forms (and their predators depend on the natural fact that the phytoplankton is not randomly dispersed.

Statistical analyses of wellplanned series of samples show

that the constituents of phytoplankton are 'overdispersed', with implies the aggregation of 'clumping' of individuals but not necessarily the forming of physically observable clusters.

Studies of a water column down to 150 metres off Bermuda, which of homogeneous that measures of temperature and salinity varied by less than 0.05°C and 0.05 per cent respectively, showed that the concentrations of phytoplankton species differed by several orders of magnitude at different depths.

There are discrete layers of phytoplankton and at times such concentrations lie near the surface, at others near or even below the depth of the euphotic layer. Members of the micro-zooplankton aggregate in these layers.

In one way or another enough individuals of any herbivorous species seem to refute Putter's conjecture, which is also true of the largest marine animals. For instance, fin whales were estimated to need zooplankton concentrations in excess of 1.5 gm per cubic metre, whereas the average standing stock of zooplankton in the Subarctic Pacific is rather less than one-tenth of this amount.

During crossings of this region Barraclough *et al.* used a 200 kHz echo-sounder and recorded shallow scattering layers that net hauls showed were composed almost entitrely of the copepod. *Calanus cristarus.* Such concentrations were considered to be more or less sufficients to satisfy the appetites of fin whales.

While there is no off season in the production of life in subtropical and tropical waters, there are seasonal changes. Indeed, monsoon comes from the Arabian word *mausium,* meaning a time or season.

During the northeast monsoon in the Indian Ocean measurements of carbon assimilated by the phytoplankton yielded a mean figure of 0.15 gmC per square metre per day, which is less than one-third of the production during the southwest monsoon. The invigorating effect of the southwest monsoon is widespread, but in the warm Pacific

Ocean seasonal changes are most pronounced in equatorial regions. Here from July to December there is greater productivily than during the rest of the year, when there are northeasterly or variable winds.

In the great subtropical gyres, where the 'seasonal' thermocline is more or less permanent nearly everywhere there can be little seasonal change in the standing stocks of plankton.

Marked seasonal changes in primary production no doubt lead to changes in the biomass and composition of the zooplankton. Concerning the productive zones of the equatorial Atlantic in the summer Bernikov *et al.* were impressed by the high biomass of plankton, the bulk of which consisted mainly of jelly-like animals, notably of pelagic tunicates.

In the autumn most of the zooplankton consisted of crustaceans. Pelagic tunicates with their efficient plant-gathering nets of mucus and relatively little living tissue to maintain, seem well organized to respond quickly to an: c ase in primary productivity.

Indeed, one slap with an instantaneous rate of population increase of 0.4 to 0.91 per day, can double its members in a single day.

Swarms of this salp, which may be very widespread, are generated immediately after phytoplankton blooms, and thus are transient resources exploited to the full. Their small relatives, the appendicularians most of which live in subtropical and tropical regions, are even livelier opportunists.

In the Black Sea the turnover time of Oikopleura is about three days. Appendicularians are among the commonest groups of zooplankton (often second only to the copepods) and, like their relatives, are able to feed on the smallest kinds of phytoplankton.

The mesh size of the filtering windows in their houses varies from species to species and when the filter becomes clogged with food, another house is quickly secreted (which may be several times in one day). In each discarbed house is a concentrated package of phytoplankton, detritus and mucus, food for other planktonic animals and bacteria.

Appendicularians themselves are prey to such animals as jelly-fishes, siphonophores, arrow-worms and lantern-fishes. By-products of zooplankters, particularly of appendicularians, salps and pteropods. form many of the flakes and larger organic aggregates of the 'marine snow' so often seen in the open ocean by divers and observers in submersibles.

Alice Alldredge concludes, 'These large amorphous bits of mucus and detritus serve as surfaces on which many planktonic organisms can rest or feed. Bacteria and protozoans use them as a permanent habitat.

Appendicularian houses and other particles of marine snow provide tiny solid substratas and introduce heterogeneity into an environment that is generally considered to be relatively homogeneous pnysically.

Although these miniature habitats have not been completely explored, it is certain that they have influenced the adaptation of feeding strategies of many planktonic organisms'. Pelagic tunicates are what theoretical ecologists call r-strategists, opportunistic animals that grow and reproduce quickly when conditions become favourable.

Indeed, nearly all gelatinous forms in the zooplankton, both herbivorous and carnivorous, seem to be r·strategists. Concerning the latter, medusae, siphonophores and crenophores evidently have '.. a versatile adaptation system, which facilitates flexible synchronisation of the population booms with abundance of planktonic forms which they use for food.

This synchronisation includes the capacity to vary the number of eggs; timing, rates and types of reproduction; to withstand long periods of starvation; to produce resting stages, and to form aggregations. These properties provide passive pelagic predators with the capacity of "waiting" for a favourable biotic situation to occur'.

At first sight, these gelatinous herbivores and carnivores seem to break the rule that r-strategists are small, but if large in bulk, as most species are, they are small in protoplasmic content. To find small kinds of r-strategists, one need look no further than the protozoans of the micro-zooplankton.

Compared to those of cold regions, zooplankton communities of the warm ocean have a low biomass, a high diversity with no very dominant species, a higher proportion of carnivorous species, more species with a short. life span and numerous herbivorous species with a reduced body size (but there is no such diminution in carnivorous species).

As Motoda and his colleagues say, one advantage of greater size in herbivorous zooplankters of cold waters is that they have extra space to store fats etc., to help them over the winter season.

The smaller herbivorous species of the warm ocean live faster and breed faster but store little in the way of food reserves, for phytoplankton is always available, even if the supply may be meagre, as it most often in the deep-blue gyral deserts of the subtropical belts.

In the North Pacific gyre productivity is generally limited by the quantity of nitrogenous nutrient present in the water. Indeed, nitrates are insignificant in quantity; the main source of nitrogen for plant growth seems to be ammonia and urea produced as waste products by the zooplankton and nekton.

The impoverished nature of subtropical waters, compared to equatorial and more northerl wat s of the western North Pacific, may be seen in Figure which shows the geou graphical variations in phytoplankton crops and the biomass of zooplankton in the surface waters.

In subtropical regions, where the upper waters are kept stable but impoverished by a marked thermocline, plant production barely keeps pace with herbivore consumption: there is, as Cushing calls at a quasi-steady state production cycle.

In the Indian Ocean, Timonin who analysed zooplankton communities, especially their copepod, euphasiid and cbaetognath constituents, in specific and trophic structure, contrasts regions of stable water stratification with those, in the equatorial belt where current divergences lead to upwelling and greater productivity.

Stable regions support communities with a low biomass but high species and tropic diversity whereas communities in regions of intensive divergence have a higher biomass with lower species and trophic diversity. Communities in regions of low divergence were more or less intermediate in the values of the above ecological indices.

After one organisms becomes the food of another, matter and energy are lost in the processes that transform prey tissue into predator tissue. Any such transformation can be seen to have a certain transfer efficiency, which Ricker defines as '. . . the percentage of the prey's annual production that is incorporated into the body tissues of the consumer species. This, in turn, is compounded. of two processes characterized by separate coefficients

E, the ecotrophic coefficient—the fraction of a prey species annual production that is consumed by predators (trophic referring to nutritive or food levels):

K, the growth coefficient-the predators' annual increment of weight divided by the quantity of food they have consumed.

Ricker suggests that an overall average growth coefficient for herbivores could hardly exceed 15 per cent, though the figure may be around 20 per cent for fish and other consumers of animal food. Concerning the ecotrophic coefficient of plant consumers a figure of 66 per cent seems reasonable, while for secondary consumers, such as lanternfishes which feed on zooplankton, 73 per cent is proposed.

Eventually, one may construct a simplified aquatic food pyramid such that the area of each rectangle representing a particular trophic level is proportional to the estimated production of that level. But the four levels shown in this

Figure, as Ricker points out, are given too low a trophic value. Successive steps in a food pyramid need to be distinguished from ecological levels (groups of animals of generally similar size and habit).

In order to assess an average trophic levels one has to consider within-level predation. The effect is to reduce the net biomass produced by any ecological level.

For instance, the production of lantern-fishes, which feed on zooplankton, is reduced by members of the zooplankton, such as jellyfishes and arrow-worms, that feed on larval lantern-fishes.

The overall transfer of production from one trophic level to the next is, of course, represented by the product KE, and Ricker estimates average value of about 10 per cent for the primary consumer and stage and 15 per cent for subsequent stages.

Relative productivities are thus: plants 100, herbivorous zooplankton 10, small pelageic fishes that feed on zooplankton about 1.6 and so forth. The more the links in a food chain, the very much less is the terminal productivity. In some oceanic communities, at least, Ryther considers there may be five links in the food chain.

Nanoplankton—micro-zooplankton—carnivorous crustacean zooplankton—larger carnivorous zooplankton—feeders on zooplankton—fish eaters.

But tuna, which feed, inter alia, on lantern-fishes, saury and euphausiids, may be put between trophic levels 5 and 6. Even so, such lengthy food chains are much less productive than those in persistent upwelling areas, as off Peru and in antarctic seas.

Peruvian anchovies feed on phytoplankton as well as small forms of zooplankton, while baleen whales in antarctic regions feed on krill, which of all euphausiids comes closest to being a complete herbivore.

Anchovies may thus be given a trophic level of about 1.5 : whales are near level 2. Ryther, who assumed that short food chains would be the more efficient, assigned *KE* values of 20 per cent to both upwelling areas, while the oceanic food chain was assessed at 10 per cent.

He concluded that the annual primary production of oceanic areas was generally low and gave 50 gm C/m^2/ year as an average figure. The corresponding figure for the Peruvian area was estimated as 300 gmCJm^2/year.

Such estimates are provisional and may well need rev ision. For instance, Cushing, after analysing data from the International Indian Ocean Expedition, concluded that the transfer of living material from primary to secondary trophic levels is three times as efficient in the open ocean as in upwelling areas.

Perhaps there are less than five links in the oceanic food chain. If there are four rather than five, then the productivity of planktivorous fishes and piscivorous tuna will be'tentimes greater than estimated.

There is still much uncertainty. As Rothschild argues, one could take a number of plausible alternative values for the trophic coefficients and arrive at very different conclusions.

Mesopelagic Food Webs

The open ocean, which extends over nine-tenths of the entire oceanic area, has a total; roductivity that is more or less equivalent to a plant assimilation of about 15 thousand million tons of carbon per year.

Investigations in the Sargasso Sea have led Riley to estimate that about 75 per cent of such production is consumed by zooplankton and heterotrophic organisms in the uppermost part of the water column.

Twenty per cent of the total production falls to the animals and heterotrophs of mesopelagic levels, the bathypelagic zone and the bottom, Mesopelagic metabolism down to and including the level of the oxygen minimum layer accounts for most of the deep-sea consumption of organic matter, which leaves less than 5 per cent to the bathypelagic and benthic organisms.

Riley suggests that total amounts consumed in the

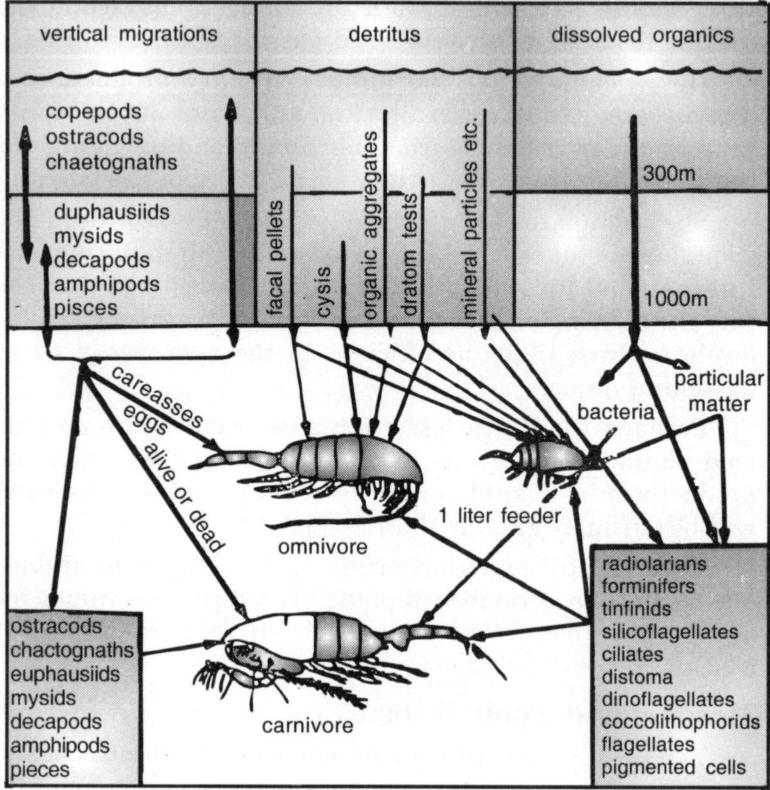

Figure 9.1: Reconstruction of the possible and likely food sources of filter-feeding, omnivorous and carnivorous deep-sea copepods.

bathypelagic and benthic zones are probably about equal. In the former, consumption is thinly (but unevenly) spread over a water column with a mean depth range of some 3,000 metres, whereas benthic consumption is concentrated into a narrow depth range in, on and just over the deep-sea floor.

In temperate and subarctic regions of the Pacific, which are something like three times are productive as the waters of the Sargasso Sea, Bogdanov estimated that 10 per cent of the organic matter produced in the euphotic zone reaches depths below 3,000 metres, but in tropical and south temperate latitudes the corresponding figure is less than 5 per cent. Clearly, there are certain outstanding questions. How is organic material transported from the euphotic zone to the depths ?

What are the metabolic requirements of deep-sea animals and how are they met ? Concerning the first question, there is the gravitational settling of particulate mutter and the movements of animals, whether daily, seasonal or ontogenetic, up and down the water column.

Gravitational Transport

When particulate material in sea water is collected on a membrane filter, or in a settling cylinder, microscopic scrutiny always reveals the presence of many particles of irregular shapes and sizes. These particles, known collectively as detritus, consist of organic and inorganic material.

Organic particles (the conentrations of which are expressed in amounts of particulate organic carbon—POC), are derived largely from metabolic waste the dissolution of pelagic organisms, and from dissolved organic matter drawn out of solution to form organic aggregates. Coacerning the first and third sources, Sieburth under a section entitled 'Fecal Fragments', writes:

'It is obvious that much of the amorphous and flake types of suspended debris occurring throughout the water column is the debris resulting from feeding, The increase in suspended organic debris following phytoplankton blooms led to the concept of organic aggregates in which large reserves of DOC formed flakes and amorphous particles through a foamtower or bubble-scavenging mechanism or through film collapse.

A reevaluation of bubble-salvaging indicated that it could be a laboratory artifact. It has been shown that particle formation is mainly in the <4,am range and is due to bacterial growth. This is in contrast to some 90% of the particles, which are in the 8-44,um size range.

'Not all feces or their fragments are consumed by coprophagy. Just as there is a threshold level for DOC utilization there is also a minimum threshold level for particulates below which filter-feeding organisms will not feed. This ensures a more or less steady state input of debris into the sea floor.

This material as well as larger debris undergoes biodeterioration on the sea floor to form a secondary production for the development of benthic ecosystems.'

The concentrations of both dissolved and particulate organic matter vary widely in the upper waters, but at depths below 200-500 metres there is relatively little change over most of the ocean. Deep DOC ranges from 0.5 to 0.8 mgC/litre, while the corresponding figure for particulate organic carbon is 3-10 μgC/litre.

Thus, DOC is hundreds of times the more plentiful, though nearly all of it seems to be unusable by living organisms. How nutritious, then, is particulate organic matter?

When, for instance, one looks in the house of an appendicularian, the gut of a copepod or the filtering-basket of a euphausiid, fragments of detritus are discernible. Deep-sea detritus, as histochemical tests reveal, contains proteins and carbohydrates, some of which should be digestible.

But laboratory tests have at best provided conflicting evidence that zooplankters, such as copepods, thrive on a detrital diet. This seems curious in view of Gordon's finding that 20 to 25 per cent of deep organic detritus is hydrolysable by proteases. There is even more direct evidence from vertical profiles of the amounts of adenosine triphosphate (ATP) in the water column.

Since ATP, the high-energy basis of metabolism, is maintained in fairly uniform concentrations in all living cells, but is soon destroyed after death, measurements of total ATP concentrations should indicate amounts of living substance.

Indeed, analyses in the laboratory show that the quantity of ATP should be multiplied by 250 to give the cellular content of organic carbon. Investigations in the eastern Pacific off California reveal that the ATP content, which is proportional to biomass, is very high in the euphotic zone but diminishes very rapidly between depths of 100 and 200 metres, below which there is a much more gradual decline to 3,000 metres.

Analyses of samples taken between 500 and 1,000 metres showed that 3 per cent or less of the particulate matter was alive but above 100 metres the live fraction ranged from 17 to 79 per cent. After concurring with other investigators that most of the dissolved and detrital carbon in deep water is unusable by micro-organisms, Holm-Hansen suggests that a small fraction of this organic carbon turns over rapidly in support of microbial populations that may well be important as the first stage in oceanic food chains.

What are the first animal stages of mesopelagic and bathypelagic food chains ? At such lev s the micro-zooplankton seems to be represented largely by foraminiferans and radiolarians; the remnants of which may be common in the guts of copepods, mysids, euphausiids and so forth.

Concerning the relatively large species of deep-sea zooplankton, the conclusions from Chindonova's analyses of stomach contents is that most of the 64 species examined, which included 15 out of 18 species of copepods, were carnivorous. But her results show also, together with Harding's more detailed studies, that some copepods feed at lower levels of the food chains.

Harding examined deep-water copepods that were taken in closing nets fished in the Sargasso Sea and slope water off Nova Scotia. The considerable range of food items that he found in about 80 species included mineral particles and detrital remains, phytoplankton protozoans, cysts and eggs, remains of cnidarians, copepods and euphausiids.

He concludes that heterotrophic protists are not only the main food of filter-feeding species of deep-sea copepods, but also form much of the diet of omnivorous and carnivorous species. In turn the heterotrophic forms '...must be supported by the dissolved and particulate organics, bacteria and possibly detrital remains of organisms present in deep waters'.

Harding's idealized chart, which outlines the food relations of filter-feeding, omnivorous and carnivorous species of deep-sea copepods and possible energy path A ays. Further studies of this kind coupled with investigations on the quantitative distribution and metabolic requirements of the main components in midwater food chains, should resolve some of the present uncertainties.

Concerning the metabolic aspect, measurements of the oxygen consumption of crustaceans and fishes from different midwater levels showed, inter alia. that species from about 1,000 metres respired at a tenth of the rate of shallow-water forms. Metabolic needs are evidently modest, but how are they met ?

Active Organismal Transport

Organic material is actively transported to the depths during daily, seasonal and ontogenetic movements of pelagic animals. Though the daily vertical migrations of aquatic animals was first recorded over 150 years ago, it is only now that there is enough direct evidence to show that migrants climb to more productive parts of the water column in search of food.

Towards nightfall, over all main regions of the ocean, myriads of planktonic and nektonic animals make their ascent, and, after spending some part of the night at higher grazing or hunting grounds, descend before daybreak to their deeper, daytime residence. In the open ocean these migrations are virtually confided to the epipelagic and mesopelagic zones.

The deepest records concern daily migrations that seem to reach a depth of 1,700 metres. Every day, then, vast quantities of planktonic and nektonic forms descend from their feeding grounds, bearing means for their maintenance and growth.

Vertical migrations '...certainly result in far greater mobilization of biological material than any other animal migration: using an average figure for plankton biomass of 25 gm/m^2 to a depth of 100 metres as a reasonable estimate for all oceans and assuming that diel migration adds 10 per cent to this at night as a minimal figure for an average daytime residence depth 250 metres deeper, then one arrives at vertical translocation of 25 tons/km^2/day over a distance of 250 metres'.

Longhurst adds that 'suth calculations are not to be taken very seriously', but his minimal figure is likely to be well exceeded in certain groups.

For instance, Baker estimates that about two-thirds by weight of adolescent and adult euphausiid populations are involved in vertical migrations between 50 and 960 metres off the Canary Islands.

In taxonomic terms, Omori states that daily vertical migration seems to occur in nearly all epipelagic and mesopelagic prawns. The same is true of mesopelagic lanternfishes. Most kinds of deep-sea migrators are cnidarians, crustaceans arrow-worms, cephalopods and fishes.

Relevant investigations on the vertical distribution of such migrators reveals that some species make extensive migration of several hundred metres, usually ending in the productive surface waters, while others are partial migrators. Members of a third group stay more or less at the same level throughout the day, but there are instances of reverse migrations.

Such categories of behaviour may also occur in individual species. At certain times some of the adults migrate while others stay still and advanced larval stages and adolescents may be partial migrators.

Concerning these and other matters, there is much to be found in the results of the Discovery Sond Cruise to an oceanic area off Fuerteventura, Canary Islands.

During vertical migrations animals are bound to experience changes in hydrostatic pressure. Over much of the ocean, particularly in the subtropical and tropical belts they are exposed to considerable changes in temperature and, if they live in an oxygen minimum layer, to wide ranges of oxygen tension.

Moreover, in seeking biological changes, they experience others, notably the attacks of predators. But their commuting seems to be timed and regulated so as to keep the light around them at a constant intensity. There is diverse evidence that light has a controlling and dominant influence on vertical migration.

In the open ocean, daily migrations are virtually confined to the euphotic and twilight zones, and the ascent of the migrators, at least, is closely related to nightfall. Investigations of the daily migrations of deep scattering layers that are recorded by echosounders, and the main sound scatterers are fishes with a swimbladder, particularly lantern-fishes, supports earlier conclusions that migrators move so as to keep the ambient light field at a constant level : they are 'imprinted' on a particular isolume.

During the daytime a given DSL will seek a higher level if the sky is darkened by clouds or a solar eclipse. Artificial lights will delay a dusk ascent or drive down a DSL by night. It would seem, then, that animals will only be able to shadow an isolume if their eyes are sensitive enough to perceive relatively small changes in light intensity.

Mesopelagic fishes have very sensitive eyes, and so have cephalopods and decapod crustaceans. But what of migrators with much simpler eyes? Concerning these and other aspects of light and vertical migrations, Cushing, Clarke, Kampa and Blaxter may be consulted.

The biological Advantages of Daily Vertical Migrations

Modes of pelagic life are other than those elsewhere. The contrasts have been illuminated by Elton who wrote '...plankton has no cover structure in which predator hunts for prey, though in most other communities it seems quite likely that without such cover the community would actually collapse before long, I believe that for this reason plankton is profoundly different in its organization from terrestrial and most aquatic communities. Two features can be noted, though the whole subject merits much deeper comparative attention.

For one thing, plankton has rather few species of animals living together; for another, its survival see-is to necessitate very intricate vertical movement.' Perhaps Elton was thinking of temperate or cold seas when writing about the 'rather few species' in the zooplankton.

A qualification is needed to cover the much greater diversity of plankton in tropical regions, but this simply means that the 'whole subject' needs even 'deeper comparative attention'.

The biological significance of daily vertical migrations is not to be considered in isolation. An outstanding question is this: Why have some species taken to such vertical commuting seemingly to make a living, while other species, which may be near neighbours of the migrators, habitually stay in the depths?

It would seem that one needs the concepts of ecological niche and competitive exclusion. A niche comprises the space and resources involved in a species' way of life. 'Niche breadth is simply the sum total of the variety of different resources exploited by an organismic unit'. Competitive exclusion means that no two species can occopy the very same niche.

Investigations during the Discovery Sond Cruise, which centred on the vertical distribution and migrations of deep-sea animals in the upper 950 metres of a restricted oceanic area, close to the island of Fuerteve-ntura are again particularly relevant.

Vertical patterns of euphausiids make a suggestive basis for comparison with other taxa, especially in view of Baker's conclusion that in principally non-migrant genera there is a decided teddency for congeneric species to occupy separate strata in the water column, whereas migrants, such as Euphausla spp., have overlapping, or even coincident, vertical ranges.

Recent trials in the Western North Atlantic with a multiple opening closing net have yielded samples that give strong support to Baker's generalization. Now it is evident from Roger's work in the equational Pacific that migrating euphausiids feed principally and most actively by night, If the same is true of the Canaries region, then it seems natural that migrating species should occupy overlapping or co-incident vertical ranges by day.

They are hardly competing for food, but merely for non-trophic living space, and evidently the different species have evolved compatible arrangements. If they feed at discrete subsurface levels by night then interspecific competition will again be minimized.

To obtain evidence of such closer-packed nocturnal segregation requires a carefully planned series of hauls, but there is some indication from Roger's investigations.

For instance, in the genus *Thysanopoda,* two species, *tricuspidata* and *equalis,* migrate to levels above the thermocline, while the populations of three others, *pectinata, monacantha* and *orientatis,* though also found above the thermocline, are concentrated mainly between 160 and 300 metres at night.

But even if *tricuspidata* and *aequalis* feed at the very same levels, they will not be competing, for the first is largely a herbivore, while the second is a carnivore. The three other species reach much the size and all are essentially carnivorous, and, as Brinton found, have nearly the same distributional ranges.

It seems reasonable that two or more related species will avoid competition if differences are evolved in their size, diet, feeding places and feeding periods. There is no evidence from Roger's study that the three above *Thysanopoda* have segregated feeding periods, but there is plenty of space, say between 100 and 300 metres, for the occupation of distinct feeding levels. Further investigation is needed.

In the Canaries region non-migrant euphausiids seem

to be separated by a middle mesopelagic association of migrant species into an upper group of segregated species of *Stylocheiron* and a lower (mainly lower mesopelagic) group consisting of five species belonging to three genera. Concerning the last group, Baker states that *Thysanoessa gregaria* and the typical form of *T. part a occupy* different depth ranges, with the setose form of *T. parva* lying in between.

There is also evidence from Rogers' observations that non-migrant euphausiids are less closely tied to nocturnal feeding periods than are migrants. Indeed, *Stylocheiron spp.,* most of which range from epipelagic to upper mesopelagic levels, are day time feeders.

Migrating and non-migrating euphausiids differ trenchantly in structural plan. The migrators, such as species of *Euphausia,* and *Thysanopoda,* are generalized forms withh spherical, undivided eyes and rather uniform, coarsely setose, comb-like thoracic limbs, designed to sweep or strain food from the sea.

These species tend to be omnivorous, but at one end of the trophic scale there are mainly herbivorous types, at the other preferential carnivores.

All the non-migrators belong to more specialized genera that have a conspicuously elongated, anterior pair of thoracic legs, behind which is a much-reduced thorocic basket.

In the most diverse genus, *Stylocheiron,* each eye is formed of a distinct upper and lower lobe, and the elongated legs are more or less chelate.

These euphausiids have definite carnivorous tendencies and they are generally much less abundant than the predominantly herbivorous kinds of migrators.

Migrating and non-migrating euphausiids are also generalists and specialists in other respects. Consider a subtropical or tropical species that migrates between 400 and 50 metres.

Besides the possession of an adaptive organization, particularly biochemical, to tolerate considerable changes in pressure temperature and oxygen tension the species must also have the right set of physiological and bebavioural attributes to spend a divided life in two quite distinct environments.

It is surely a generalist. The specialist that stay at one level display no such ranging adaptablity. As regards the population structure in the water column, the generalist migrators are certainly eurytopic, the specialist stenotopic.

Are there the same differences in the horizontal plane ? If the two large genera, *Euphausia* and *Stylocheiron*, are compared, and both consist predominatly of epipelagic to upper mesopelagic species, it emerges that species of the former cover a much wider area.

Euphausla is represented in the substractic, antarctic and subarctic zones and in the subtropical and tropical belts while *Styl cheiron* is essentially a warm-water genus.

Thus, evolutionary potentiality in the genus *Euphausla* has been such that species have emerged that are not only adapted to the wide range of physical and biological conditions associated with diel vertical migrations in the warm ocean, but also, inter alia, to the widely ranging seasonal conditions, especially biological, in colder regions.

One is reminded of the recent work of Jackson, (who finds that the infaunal bivalves adapted to the rigours of life in depths of 1 metre or less have distinctly wider geographic ranges than other infaunal species restricted to depths greater than 1 metre. Again, the set of species with wider environmental tolerances has the greater geographic representation.

Sergestid prawns, some of which prey on euphausiids, are active migrators. Off Fuerteventura, Faxton found that species of the subgenus *Sergeste,* which are semi-transparent and bear internal, ventral light organs, live at levels above 750 metres, whereas *Sergia* spp., uniformly red and with external light organs, live below this level.

Here, in a daytime descent from 400 metres and considering only the migrators, there are three pairs of species that each live together in concentrations between the same levels of the water column ; *Sergestes sargassi* and *S. armatus S. vigilax* and *S. pectinatus* and *Sergia robustus* and *S. splendens.*

There is also *segestes corniculum.* Partly or entirely below the last pair, there are two non-migrating bathypelagic species, *Sergia japonicus and S. tenuiremis.*

For investigations on both the vertical pattern and food of sergestids, we turn to the Ocean Acre project off Bermuda, During the day there is again a step-down occupation of the water column with groups of two or more migrant species concentrated between the same levels: *Sergestes cornutus* and *S. sargassi: S. pectinatus, S. vigilax* and *S. armatus, S. atlanticus S. corniculum* and *S. grandis* and *Sergia robustus* and *S. splendens.*

Again there are the same bathypelagic species, *Sergia japonicus* and *S. tenuiremis,* but in the Bermuda area they appear to be partial migrators.

During both day and night the only species that occurred together were *Sergestes sargassi, S. pectinatus* and *S. vigilax,* but dissimilar features of their third maxillipeds, a key pair of limbs involved in feeding, led Donaldson to suggest that they have different diets.

Donaldson found that most species took a variety of food organsims, but three species contained euphausiids more often than other kinds of prey.

Fish were most frequently found in *Sergia robustus* and *grandis.* The bathypelagic species, *Sergia japonicus,* which has a flaccid body with reduced musculature, appears to be a detritus-feeder.

In most species food items were found most often at night, but no species was exclusively a nocturnal feeder. On the other side of the Atlantic, however, Foxton and Roe who analysed the food of **six** sergestid and three oplophorid prawns, concluded that intensive night-time feeding seemed to be the rule in all nine species.

Once more, partitioning of resources and the water

column by a set of migrating species seems to be such as to reduce interspecific competition. During the daytime in particular, groups of two or more species may coexist in relatively narrow strata, which is quite unlike the mutually exclusive stratification in the non-migrating euphausiid, *Stylocheiron.*

How far do copepods show the same trends? They are the prey of innumerable marine animals from cnidarians to whales (including) euphausiids and sergestids) and at any one point in the warmer parts of the ocean, many species occupy the water column from top to bottom..

In the Fuerteventura area 212 species of calanoid copepods were identified in hauls made between 960 and 40 metres.

The bulk of the population, which consisted largely of immature stages, was centred at 5k;0 metres by both day and night. Of the adult forms, some were conspicuous migrators but most species showed restricted migratory tendencies or were non-migrators.

Considering first the non-migrators, it is evident that related species do not necessarily follow the discretely stratified. pattern. For instance, three pairs of species: *Calanus belgolandicus* and *C. carinatus; Rhincalanus nasulus* and *R. cornutus* and *Spinocalanus magnus* and *S. spinosus* were concentrated between much the same depth ranges throughout the entire day.

After the name of each species in parentheses is the length range of the females and it will be seen that in each pair one species is decidedly the larger. The significance of size differences in reducing competition between coexisting, congeneric species of animals has been well demostrated by Hutchinson who found that the ratios of mouthpart sizes in relevant species of insects, birds and mammals ranged from 1.2 to 1.4.

As Wilson observes, larger predators are able to utilize food sizes that are not available to smaller predators, while the reverse is much less true. Within a species competition for food may be reduced by sexual differences in size but in most oceanic copepods the males are slightly the smaller sex.

On the other hand, the adult males of some species take no food. There is every indication, then, that Hutchinsons's rule also applies to certain coexisting, congeneric species of oceanic copepods. Species of *Undeuchaeta andPleuromamma* are among the conspicuous migrators. By day *Undeuchaeta major* and *U. plumosa* were concentrated between 450 and *900* metres by night the migrators had reached levels above 250 metres.

But if most of their food is taken at night there need be little competition, for once more, the size difference between the females (in parenthesis) is considerable. Two common species of *Pleuromamma, xiphias* and *abdominalis* had overlapping distributions by day, the first centred at 570 metres, the second at 400 metres.

By night most of the extensive migrators were clustered around levels of 250 and 100 metres, but yet again, size differances are such as to reduce competition. Moreover, in the uppermost waters Roe has evidence to suggest that *x1phias* was most abundant between 50 and 25 metres, *abdominalis* between 10 and 0 metres.

Lastly, three other, species, *borealis piseket gracilis,* are centred at 350, 300 and 200 metres, respectively, during the day; by night the first is concentrated at 250 and 100 metres, the other two at 50 metres. Nocturnal competition between *piseket* and *gracilis* is possible. They are of much the same size and all three species, as Roe observes, are closely related.

He adds also that samples from the uppermost 50 metres showed that *borealis* was scarce and *pisekel* more abundant than *gracilis.* Further investigation should be of interest. The evolutionary interplay of mesopelagic life seems to have been such the potential competitors in the zooplankton and micronekton have come to be more or less segregated through mutually exclusive combinations of spatial, temporal and biological attributes.

The same is true of the productive upper waters, as may be seen in Zalkina's (1970) paper on the vertical distribution of cyciopoid copepods in the equatorial western Pacific. During a 24-hour station a series of hauls was made between the surface and 300 metres. The catches showed that the species of Oithonidae, which are non-migrators and concentrated between 75 and 150 metres, have little overlap in the water column with those of the families Oncaeidae and Corycaeidae.

It is interesting also to compare the changing vertical dispositions of the two dominant species, *Olthona* ***plum jera*** and ***Oncaea media.***

Again, there is a little overlap, except from 10.00 to 18.00 hours, when the second species is keeping its daytime station. Zalkina found that both migrating and non-migrating species tended to maintain a step-like distribution. Some species differed in the times of their ascent and duration of stay at higher feeding levels, and Zalkina concluded, 'This peculiarity of disposition tends to lessen the intensity of food competition between species with similar nutrition habits.

'She also quotes her colleagues, particularly M.E. Vinogradov, who also found that dominant species of oceanic zooplankton with similar food requirements occupy discrete strata in the water column, and that such segregation is particularly marked in regions of low productivity.

The Californian sardine is an active fish, and during its virtually restless life, respiration consumes most (82.99 per cent) of the calorific content of its hard-won food in the zooplankton. Reproductive processes take a mere I per cent of assimilated calories.

The lives of mesopelagic fishes are not so strenuous, but whether they migrate or not, they are highly organized predators. needing the means to run elaborate tissue systems, particularly the substantial muscles of their jaws and motor complex. How do they share available resources?

Most kinds of mesopelagic fishes are lantern-fishes,

alepisauroids or stomiatoids. Of all the twelve families that form these groups, daily vertical migrations are by far the most prevalent in the lantern-fishes.

Indeed, except for some of the deepest-living kinds, such as certain *Lampadena* species and those of the genus *Taaningichthys, well* planned investigations reveal that lantern-fishes are inveterate migrators. Judged by their biomass, ubiquity and species diversity, they are also the most successful family of midwater fishes.

In relative length and other features of their jaws, which bear rows of small conical teeth, there are three main groups ; small-jawed, medium-jawed and long-jawed. The genus *Diaphus* is unusual in having representatives in all three groups.

One species, *Diaphus taaningi,* a rather small lantern-fish with medium-sized jaws, is known from the Gulf of Mexico, the Caribbean Sea and off West Africa. In the Caribbean it lives in the Cariaco Basin, a depression in the continental shelf off Venezuela with a shallow northern still that restricts the horizontal exchange of water between Basin and Sea.

Due to this restricted circulation and other factors the Cariaco Basin is anoxic below a depth of about 300 metres. But *D. taaningi* lives above the anoxic waters. Baird and his colleagues found that it migrates from a daytime depth of about 250 metres to appear in the upper 50 metres at night.

Nearly all of their fish ranged from 38 to 45 mm in standard length, and all were feeding largely on herbivorous zooplankters, particularly copepods and larvaceans, which formed about 70 per cent of the total prey biomass.

Analyses suggested that most of the prey was taken in the early evening, leaving the rest of the night for digestion. ' Fish taken before the evening ascent were essentially empty. In their discussion the authors argue that *Diaphus taaningi* and certain other midwater fishes, such as the hatchet-fish, *Polydpnus-* are bound to near-neritic waters where there is an abundance of small zooplankton.

As prey density and distribution at a given time are thought to determine the amount of food a predator can obtain as well as the cost to the species in energy of pursuit, it would appear that small prey in high density in the night time depth zone may well be a niche requirement for *D. taaningi*.

Two other lantern-fishes, *Stenobrachius Icucopsarus* and *Triphoturus mexicanus,* which are long-jawed and larger than *D. taaningi* were the dominant midwater fishes taken during a cruise off California. Collard, who found that both preyed mainly on small and medium-sized crustaceans, particularly copepods and euphausiids, concluded that diets were diverse, with no sign of selection.

But some mesopelagic fishes feed selectively, as Merrett and Roe (1974) found after analyses of fish caught by *Discovery* in a series of eleven hauls over a 24 hour period between 230 and 266 metres in the eastern North Atlantic.

By comparing the composition of the zooplankton (in the net) with fish prey, the authors were able to determine whether or not food items had been selected by some of the more abundant fishes. They looked also for periodicities in feeding. The small stomiatoid *Valenciennellus tripunctulatus,* virtually a non-migrator, fed mainly by day on selected calanoid copepods.

Two partial migrators, the hatchet-fishes, *Argyropelecus hemigymnus* and *A. aculeatus,* which seemed to feed mainly at dusk, differed in diet. The first fed randomly on mixed copepods and ostracods, the second selectivity on ostracods. There were parallel results from two fully migrant lantern-fishes, *Lobianchia dofleini* and *Lampanyctus cuprarius.* Both fed at night, the first randomly on copepods, the second on selected amphipods, particularly the males of *Phrosfna semilunata.*

The viper-fish *Chaullodus danae,* an avid predator, seemed to feed at any time and randomly. Merrett and Roe concluded that vertical migrations and feeding habits are closely connected. whether by the migration of prey species away from levels occupied by their predators or by the upward migration of the predators to the headquarters of their prey.

At the same station another series of nets was fished down to 2,000 metres, and from these catches Badcock and Merrett have presented an overall analysis of vertical patterns in the midwater fish fauna. Considering first the lantern-fishes, their results complement those of Badcock from the Fuerteventura area off the Canary Islands.

All species, or at least all wellrepresented species, were diel migrators. By day the vertical ranges of silvery species largely overlap or are coincident between levels of 300 and 700 metres, while dark species have similar range correspondences below 700 metres.

Silvery lantern-fishes and transparent species of sergestid prawns have much the same depth range : so do dark lantern-fishes and red sergestid prawns. Indeed, lantern-fishes and migrating species of sergestid prawns and euphausiids coexist at similar or widely overlapping depth ranges by day. They also tend to take most of their food by night. There is also evidence of dietary differences amongst them.

Over most regions of the ocean between the subarctic and subanta-rctic zones the most numerous non-migrators at mesopelagic levels are the more or less transparent species of *Cyclothone,* stomiatoid fishes known as bristle-mouths.

Dark *Cyclothone* live deeper and like other bathypelagic animals, which are considered later, are also non-migrators. The transparency of mesopelagic bristle-mouths is interrupted by black markings, especially at internal levels, but in *Cyclothonc pscudopallida* there is also an integumentary dusting of small melanophores over much of the body.

When adult they have a gas-filled swimbladder, which regresses and becomes fat-invested in their dark relatives. Further differences between the two groups are considered elsewhere. At mesopelagic levels in the Canary Islands region there are two 'transparent' *Cyclothone, braueri* and *pseudopallida*.

There is also a third species *pallida,* whieh ranges from lower mesopelagic to bathypelagic reaches. The first is concentrated between 400 and 600 metres, the second between 500 and 800 metres and the third between 600 and 800 metres. There is thus relatively little overlap between *braueri* and the other two species and *braueri* is not only the dominant species but also the smallest (maximum standard length 38 mm v. 45 mm and 63 mm).

Competition between *braueri* and the deeper-living pair is evidently slight and is reduced also between mature males and females of *braueri,* to consider only the dominant species. As in other *Cyclothone,* the males are not only the smaller sex but nearly all live at higher levels than the maturing to ripe females.

Apart from reducing intraspecific competition, the segregation of the sexes may well facilitate eventual sexual congress. Competition between mature individuals of *pseudopallida* and *pallida is* likely to be modulated by their difference in size, but other information is needed, particularly on their diet and trophic periodicity.

The concentrations of these two species also overlap with the upper part of the depth range of *Cyclothone microdon* but this is made largely of young individuals which are not likely to compete much with mature fishes of the other two species.

Lastly it will be recalled that Badcock and Merrett's paper is based on sampling at 30°N, 23°W, about 600 miles west of Fuerteventura, Canary Islands. At this station *Cyclothone microdon* is a dominant species, but off Fuerteventura it was not recorded.

Faunal composition in the water column changes from place to place but there is still the interplay of non-migrating species to reduce competition. Off Fuerteventura the relations between *C. braueri. pseudopallida* and *pallida* are rather like those at the westerly station, though there is less overlap between the second and third species, and *microdon is* replaced as it were, by two other dark-coloured bathypelagic species, *C. acclinidens* and *livida.*

Seven of the nine dominant species of lantern-fishes in the Bermuda area are small or diminutive, about the same size as the mesopelagic species of *Cyclothone.* Like *Diaphus taaningi* these lantern-fishes must be restricted by their relatively small gapes to small prey organisms.

If any one of the seven is compared to a *Cyclothone* of the same length, the latter will be seen to have a considerably wider gape, for the jaws are about four-fifths the length of the head, which is about a fifth to a quarter of the fishes standard length.

The *Cyclothone* should be able to take a wider size range of prey items than can a diminutive lantern-fish. But the latter migrates up to epipelagic levels or just below, where there is likely to be a greater abundance of small prey, particularly of herbivorous copepods, both young and adult.

To make a living at deep-sea levels, where, with increasing depth, there is a more or less exponential decline in the biomass of suitable food, it would seem that a non-migratory fish should be adapted to secure the widest possible size range of food organisms.

Indeed, investigations off southern California showed that *Cyclothone acclinidens,* which lives below the transparent species, *C. signata,* at depths greater than 400 metres, takes a wider size range of prey. from copepods to arrow-worms and amphipods.

This is understandable, considering that the deeper-living species bears the larger jaws, due both to its greater size (maximum standard length 50 mm v. 33 mm) and to its relatively larger head.

Similarly. Ebeling and Cailliet found that the deepestliving and non-migratory fishes called big scales not only have larger jaws than shallower-living species, but also develop a gill raker system fine enough to retain and concentrate food items as small as 1 mm or so.

The larger a fish, the less energy it needs per unit of body weight. Even so, a large fish cannot exist in a food-poor environment : there are no deep-sea tunas and billfishes. Some of the largest mesopelagic fishes are alepisauroids, but unlike their relatives the lantern fishes, they are nearly all nonmigratory.

The lancet-fish *Alepisaurus ferox* grows to about 2 metres, but it is not a heavy fish. A five-foot individual weighed about 4 pounds. which is not only indicative of its long compressed body but especially of its lightly ossified skeleton and sparely developed motor muscles.

It is above all due to the reduction of skeletal and muscular systems that deep-sea midwater fishes nearly attain neutral buoyancy.

Thus, as well as being economically made, they need little energy to keep to a preferred level in the water column. The barracudinas, which range in length from a few to over 50 cm, have both the look and economical build of their relatives, the lancet-fish : so have other alepisauroids.

These fishes have also the jaws, swallow and maw to take very large meals. An individual of one species, *Anotopterus pharao,* which was swallowed by a whale captured in the Ross Sea, Antarctica, and measured nearly 75 cm, contained two barracudinas of 18 and 27 cm in length. The barracudinas were bloated with krill, *Euphausia superba.*

More recently, Wassersug and Johnson (1976) have shown that alepisauroids have a sac-like stomach that is both large and greatly distensible. The intestine is simple and straight.

Species of all six families swallow intact prey that are at least one-third to one-quarter of their length. As the two authors say, in energetic terms midwater predators in the deep sea must not be much larger than their prey. But given the above and other adaptations alepisauroids can afford, at it were, to be non-migrators. But there are non-migratory, tiny-mouthed mesopelagic fishes. Such argentinoids have large upturned tubular eyes with parallel optical axes, which, together may well provide extra sensitivity.

Even more important, such eyes presumably afford much-needed accuracy in pin-pointing small prey that is to be sieved in tiny jaws.

Hatchet-fishes of the genus *Argyropelecus,* which at most are partial migrators, have similar eyes but larger jaws. Even so, accurate prey taking is still needed, particularly when the diet is selected, as in *A. aculeatus.* Where prey of a suitable size may be sparse, chances of nourishment should not be missed. And there is a further aspect.

The hatchet-fish *Argyropelecus affinis* and the alepisauroid *Scopelarchus analis, a* voracious fish, also with tubular eyes, have yellow lenses. The function of such lenses may well be to enable their owners to 'see through' the bioluminescent camouflage of various midwater animals.

Meals may be seen that might be missed by predators without special filters. More evidence could be adduced to support the idea that diverse mesopelagic fishes have each evolved a unique and special organization to enable them to stay in the depths and make a living after their metamorphosis to adult form.

The same is true of non-migratory *euphausiids. Migrators* are generalists, adaptable to wide ranges of physical and biological conditions not only vertically but also in the horizontal plane. Is this also true of fishes? For instance, one can compare the distribution of the migratory lantern-fishes with their nonmigratory relatives, the *alepisauroids.*

To take but one criterion of extensive distribution, about 15 species of lantern-fishes are known from the antarctic zone but only three or four alepisauroids. Daily vertical migrations, evidently in search of food. are thus ingrained in many kinds of *epipelagic* and mesopelagic animals. One advantage is that during their resting, non-feeding periods, potentially competing species can reside together between much the same depth levels.

Indeed, the entire nexus of vertical migrations would seem to be *fundamental* to species packing and community structure in the water column.

Both these aspects are well illustrated by fishes, particularly as they are evidently the outstanding components of the deep scattering layers recorded on ships' *echosounders.* The prominence of fishes as acoustic scatterers is due largely to their gas-filled swimbladder, which is present in most lantern-fishes and other *mesopelagic* forms, notably in the smaller *stomiatoid* fishes.

Some scattering layers undergo daily vertical migrations, but others are more or less *stationary.* In the present context a ranging survey of DSLs in the North Atlantic by Haigh (1970) will be considered. Records of a 10 kHz echo-sounder from 1963 to 1967 and between latitudes 10° and 68°N revealed five *mesopelagic* layers during the day.

The first three (ranges of maximum mid-layer depths: A 135-170 fms, B 140-245 fms, C 215—280 fms) were purely migratory and found in latitudes south of 49°N. Layer D, which was partially migratory and attained a maximum depth of about 330 fms, was the most widespread, being *recorded* at all latitudes. Layer E. was mainly *migratory* and recorded only in restricted areas in latitudes south of *Iceland.*

Thus, north of 49°N there was one main layer D joined locally by layer E. South of this latitude there were up to five layers and over wide areas three occurred together. The conjecture is that the number of DSLs is a reflection of species diversity, which is much greater under the *armor* waters south of ca. 40°N than in the colder waters to the north. Where there are more species there is need for more statified *speciespacking* in the water column.

Indeed, considering, for instance, that there are over 200 species of mid-water fishes off Bermula it is remarkable that there are so few deep scattering layers. That there are so few would seem to be related to the phenomenon of daily vertical migrations, whereby sets of potentially competing species can live together between the same depth ranges when they are resting by day.

During the might, all but the non-migratory and partially migratory layers are in the upper 150 metres, where broad and prominent sound scattering is recorded. The biomass and number of species in a DSL maybe greater than those outside but do the sound scatterers and associated species form *communities*?

Off Santa Barbara, southern California, factor analyses of the catches of deep *midwater* animals revealed a complex of four resident *communities*, which interact with a number of *transitory*, *presumably* less stable, groups. On a larger scale there are the investigations of Backus and his colleagues in the *Atlantic*.

Daylight hauls were generally made at the depth of sound scattering maxima, while might hauls were set to sample concentrations of migrating fishes at depths above 300 metres. Though largely concened to sample layers of strong sound scattering, the investigators are readily able to refer the distributions of the mesopelagic fishes in their catches to six or more pelagic regions, each with a ' ... more or less unique fish fauna with its characteristic assemblage of species in *characteristic* proportion, its characteristic diversity and so on'.

This definition could also be applied to the communities that presumably contribute largely to the individual characteristics of a pelagic region. In turn, the more or less stable structure of a community is integrated by the many relationships between its constituent species (particularly predator-prey interactions and interspecic competition) and the environment.

With such community criteria in mind, it will be interesting to see how far they apply to the animals in sonic scattering layers. There is much to do. As Kinzer says, we know little about the predator-prey *interactions* of organisms within the *diurnal range* of such layers.

If daily vertical migrations are feeding sorties, the metabolic benefits must exceed the energetic costs of ascent and descent. No doubt the benefits will vary from day to day, but even so, the energy needed for vertical migrations by small crustaceans is but a few per cent of their daily metabolic rate.

For *planktonic crustaceans* ranging from 0.5 to 30 mm in length, the amount of energy expended during vertical migrations was calculated (by estimating the work required to over come hydrodynamic resistance and lift their net body weight) to be no more than 20-40 per cent of their resting metabolism. *Comparable* figures for small fishes may well be of the same order.

The adaptive value of daily vertical migrations in energetic terms has been considered by McLaren. In a study of migrating zooplankton McLaren followed the implications of a mathematical model based on von Bertalanffy's growth equations.

The model predicts that a *migration* from cool to warmer waters can be advantageous : an animal that is mainly active when feeding at upper levels and then descends to rest in cooler waters, will use its food to good effect. For instance, growth and fecundity should be enhanced, for during its resting periods an animal should eventually grow larger than it would in warmer surroundings.

Though subsequent tests of this hypothesis have not been encouraging, it should not yet be abandoned. Besides the gain in food, the lessening of competition and the partitioning of resources, there are other likely adaptive aspects of vertical migrations.

For planktonic animals, at least, Hardy, after pointing out that during their migrations currents may carry them considerable distances in the horizontal plane, suggests that, '... the development of a habit of regular periodic vertical migration has been evolved to enable the animal continually to sample new environments- new feeding grounds and so to have some power of choice, of varying its behaviour within limits'.

Following this line of thought, *Isaacs*, Tout and Wick have shown how animals may be transported into areas of high standing crops of phytoplankton.

Besides gaining food, vertical migrators may escape predators. For instance, in the *tropical Pacific yellow* fin tuna and albacore *evidently* feed during the day, when lantern-fishes and other likely prey are residing at lower levels.

Though one cannot imagine lantern. f shes with their sensitive *nocturnal* eyes in the surface waters by day, it seems also that they avoid lancet-fishes, voracious mesopelagic predators. In both the North Atlantic and the southeastern Pacific the fish prey of *Alepisaurus* consists largely of *hatchetfishes*, *barracudinas* and others of their kind.

All such prey are non-migrators ; lantern-fishes, nearly all of which migrate, are rarely taken. Clarke also observes that studies of the feeding habits of tuna, dolphin-fishes (Cor} *phaena*) and *Alepisaurus*, indicate that they consume very few vertically migrating forms such as lantern-fishes and gonostomatid fishes.

Thinking of *epipelagic* predators, particularly yellow fin tuna and *albacore*, *Roger* and *Grandperrin* argue that if such predators do not feed at night, they cannot retrieve the energy won by the migrating mesopelagic fauna.

They see the whole process as an 'energy valve' permitting the downward transfer of energy but reducing the upward one. They conclude that if this process exists in other parts of the food chain ' , . . then one can suggest that the mechanism is fundamental, as it explains how the deeper layers of the ocean are able to support a *fairly large biomass*'. Their ideas seem well worth pursuing. Lastly, ontogenetic and seasonal migrations also lead to transfers of organic matter between the *euphotic zone* and *mid water levels*. Concerning the former, we shall see that in many kinds of mesopelagic and bathypelagic animals the spawning place and buoyant properties of the eggs are such that the larval life history takes place in the productive surface waters. During aid *metamorphosis* there is a descent to the adult-living space.

There must be heavy mortality as the eggs and early larvae ascend to their *euphotic nursery* ground and during their stay in this layer. The descending metamorphosis stages will also fall to *predators*, but the young are now large and active enough to avoid some of their *enemies*.

INDEX

A

B

C

D

E

F

G

H

I

N

O

P

Q

R

S

T

U

V

W

X

Z